APPLICATIONS OF
OPERATIONAL
AMPLIFIERS

THE **BB** ELECTRONICS SERIES
BURR-BROWN

Graeme • APPLICATIONS OF OPERATIONAL AMPLIFIERS

Tobey, Graeme, Huelsman • OPERATIONAL AMPLIFIERS

APPLICATIONS OF OPERATIONAL AMPLIFIERS

Third-Generation Techniques

JERALD G. GRAEME, M.S.E.E.

Manager, Monolithic Engineering
Burr-Brown Research Corporation

McGRAW-HILL BOOK COMPANY

New York   St. Louis   San Francisco   Düsseldorf   Johannesburg
Kuala Lumpur   London   Mexico   Montreal   New Delhi   Panama
Paris   São Paulo   Singapore   Sydney   Tokyo   Toronto

Library of Congress Cataloging in Publication Data

Graeme, Jerald G
    Applications of operational amplifiers.

    Includes bibliographical references.
    1. Operational amplifiers.   I. Title.
TK7871.58.06G7   621.3819'57'3   73-7984
ISBN 0-07-023890-1

The information conveyed in this book has been carefully reviewed and
believed to be accurate and reliable; however, no responsibility is assumed for the operability of any circuit diagram or inaccuracies in calcula-
tions or statements. Further, nothing herein conveys to the purchaser a
license under the patent rights of any individual or organization relating
to the subject matter described herein.

CONTENTS

PREFACE

In less than a decade the operational amplifier made the transition from analog computer subassembly to universal analog component. It did so because of its outstanding versatility in performing electronic functions. This versatile amplifier is now a basic gain element somewhat like an elegant transistor.

During this transition, the operational amplifier was applied to perform a multitude of electronic functions, and familiarity with such applications greatly simplifies the task of the analog circuit designer. By adapting appropriate applications to his specific requirements, the circuit designer is able to quickly develop precise, simplified circuits. Many of the early operational amplifier applications were made available for design reference in the Burr-Brown *Handbook of Operational Amplifier Applications*. A more recent and expanded set was published in the McGraw-Hill/Burr-Brown book *Operational Amplifiers: Design and Applications*.[1]

---

[1]G. E. Tobey, J. G. Graeme, and L. P. Huelsman, *Operational Amplifiers: Design and Applications*, McGraw-Hill Book Company, New York, 1971.

However, circuit designers continue to envision new roles for the operational amplifier, and a new generation of applications developed as the operational amplifier technology matured. This book was written to supplement the second book cited above by providing an engineering reference for the current applications, most of which were previously unpublished. Once again, an effort has been made to produce a practical reference from which a circuit designer can adapt applications. Rather than being just a collection of circuits or theoretical analyses, this book covers the operation, pertinent design equations, error sources, and relative merits of each circuit. With this information each general circuit presented can be modified for an individual requirement.

The applications of this book are divided into six chapters, beginning with general techniques for effectively using operational amplifiers, followed by general-purpose applications and then some special-purpose circuits. In Chapter 1 some of the more subtle characteristics and interrelationships of operational amplifier performance are described to aid the user in optimizing the performance of operational amplifiers. This includes means for predicting and controlling thermal drifts, gain error, and noise, as well as means for protecting against overloads. Techniques for further improvement of operational amplifier performance are presented in Chapter 2. With these techniques, it is possible to increase input impedance, isolation impedance, voltage swing, and output current. In addition, applications which produce differential outputs and true differential inputs are described.

Treatment of the electronic functions performed with operational amplifiers is given in the remaining four chapters. Chapter 3 covers some advances in integrators, differentiators, current sources, logarithmic amplifiers, multipliers, and active filters. Additional general-purpose circuits are described in Chapter 4 for signal processing functions such as comparators, multiplexers, clamps, absolute-value circuits, sample-hold circuits, and rms-to-dc converters. In Chapter 5 signal generators are presented which produce sine waves, square waves, triangle waves, ramp trains, pulse trains, and timed pulses. In most cases the characteristics of the generated signal are determined by a few components external to the operational amplifiers. More specialized circuits which perform some end function are covered in Chapter 6. These include circuits for transistor testing, measurement, control, audio, and automatic gain control. Following this chapter is a glossary of terms related to operational amplifiers and their applications.

I am particularly grateful to two individuals highly skilled in the related technology, Henry Koerner and Donald R. McGraw, for their review of the manuscript and the many resulting improvements. It is also appropriate to acknowledge considerable related information and background derived from the Burr-Brown environment over the past eight years. Additional thanks are expressed to Fran Baker for her expeditious typing of the manuscript. I also wish to thank my wife, Lola, for her accurate and attractive rendering of the illustrations.

Jerald G. Graeme

APPLICATIONS OF OPERATIONAL AMPLIFIERS

1

OPTIMIZING OPERATIONAL AMPLIFIER PERFORMANCE

Operational amplifiers perform many electronic functions in instrumentation, computation, and control. The widely diverse applications of these analog building blocks demand a similarly broad range of performance characteristics. As a result, a general specification of characteristics overlooks performance subtleties important to various applications. Furthermore, many characteristics display complex relationships with other characteristics or application conditions. Such operational amplifier behavior often defies straightforward characterization. A variety of these more subtle characteristics and complex relationships are described in this chapter to aid in optimization of operational amplifier circuits such as those discussed in the following chapters. Specifically, the characteristics described here are those often encountered in optimizing dc, ac, noise, and overload performance.

1.1 Preserving DC Performance

Other than voltage gain, the primary determinants of operational amplifier dc performance are the input characteristics. Most notably, the

1

input offset voltage, input bias currents, and input offset current and their respective thermal drifts are major sources of dc error. Application conditions affect the associated errors, but sometimes with predictable relationships which permit avoidance of further dc error. This is the case for the offset voltage thermal transient response, the offset voltage drift created by nulling the offset, the thermal drift of the input currents, and stray leakages.

1.1.1   Thermal transient response of input offset voltage

The thermal drift specified for input offset voltage characterizes the steady-state offset as a function of temperature. It does not include the transient offset variations which occur during warm-up or during rapid temperature changes. As a result, an offset voltage can easily undergo changes well above those anticipated from the specified drift. Warm-up drift is generally related to the thermal drift characteristic, as the amplifier power dissipation merely raises the temperature of the amplifier above the ambient temperature. However, if this dissipation creates nonuniform heating of the amplifier, thermal gradients which produce additional drift are established. Thermal gradients also result from rapid ambient temperature changes or from thermal shock. These gradients represent temperature differentials between matched components. Most often the greatest effect of a temperature differential occurs between matched transistors, where it disturbs the delicate but precise balance of emitter-base voltage drifts or gate-source voltage drifts in differential stages. Very little difference in temperature is needed to disturb the 3 μV/°C match of a 2,000 μV/°C emitter-base voltage drifts.

The resulting offset voltage transient responses under warm-up and thermal shock are illustrated in Fig. 1.1. Both responses place limita-

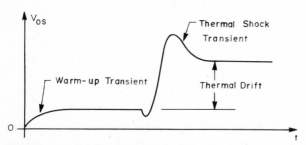

Fig. 1.1   Input offset voltage transients occurring after turnon or after a rapid temperature change can add greatly to the changes from ordinary thermal drift.

tions on the operating conditions of the operational amplifier if dc performance is to be carefully preserved.   To avoid errors from the warm-up transient, the circuit should not be used until the transient has passed. Depending upon the amplifier thermal mass, this transient usually lasts from 1 to 5 min in free air.   However, the transient is commonly equivalent to an amplifier drift of only 5 to 10 ° C, and the associated drift may not be significant in circuits intended for wide temperature range operation.   The transient response to thermal shock limits the permissible rate of change in the amplifier temperature in low-drift applications. When the rate of change exceeds the rate at which the amplifier mass can change uniformly, thermal gradients which develop errors above the anticipated drift error result.   The length of the thermal shock transient varies from 10 s to several minutes, depending upon the uniformity and size of the thermal mass.

Monolithic integrated-circuit operational amplifiers have thermal masses which are highly uniform and very small, and they provide the shortest thermal shock transient responses.   However, the thermal response characteristics of the monolithic types introduce one additional transient error.   With its high thermal conductivity and low thermal mass, the monolithic chip readily conducts heat from dissipation at its output to its sensitive input.   This thermal feedback introduces input offset voltage changes with load current variations accompanying signal swings.   In addition to creating error, these input voltage changes can override summing junction signal changes to give an appearance of infinite open-loop gain or negative output resistance.   Careful amplifier layout avoids thermal gradients caused by thermal feedback so that the thermal shock transient will only be due to the change in chip temperature.   The response then resembles the warm-up drift, which can last from several seconds to several minutes.   All the various thermal transients are reduced by heat-sinking the amplifier.

1.1.2   Drift effects of input offset voltage nulling   Almost every operational amplifier has a provision for nulling the input offset voltage.   However, most of the recommended null techniques add significant offset voltage drift.   Correction of dc error in an operational amplifier circuit can then degrade its temperature sensitivity.   A typical bipolar transistor amplifier with a 1 mV offset and $3 \mu V/°C$ drift will acquire an additional $3 \mu V/°C$ drift component when its offset is nulled.   This added drift term may have the same or the opposite direction as the original amplifier drift, so that the net offset drift can either increase or decrease.   In

fact, if the offset voltage nulled is all due to the one stage compensated, the drift is also nulled.[1] However, in almost all multistage operational amplifiers no one stage accounts for all significant offset voltages. As a result, drift will often be degraded by the offset null.

Offset voltage null is most commonly achieved by supplying a correction current to one side of a differential stage, as in Fig. 1.2. Correction current I_c unbalances the two emitter currents, as indicated, to develop a change in input offset voltage of[2]

$$\Delta V_{OS} \doteq 2I_c r_e \qquad \text{for } a \doteq 1$$

where
$$r_e \triangleq \text{transistor dynamic emitter resistance}$$

For some amount of ΔV_{OS} the net offset is reduced to zero. However, this amount of offset correction exists at only one temperature, since r_e is highly temperature-dependent. This dependence is expressed in the common equation for dynamic emitter resistance of

$$r_e \doteq \frac{KT}{qI_e}$$

where
$$K \triangleq \text{Boltzmann's constant}$$
$$T \triangleq \text{temperature in degrees Kelvin}$$
$$q \triangleq \text{electron charge}$$

Then, the offset voltage correction varies with temperature as expressed

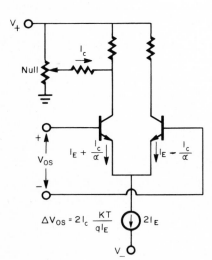

Fig. 1.2 Input offset voltage null is commonly achieved by unbalancing the currents in a differential stage.

by

$$\Delta V_{OS} = 2I_c \frac{KT}{qI_e}$$

Consideration of the junction equation reveals the associated change in thermal drift for a bipolar transistor input stage to be[3]

$$\frac{d(\Delta V_{OS})}{dT} = \frac{\Delta V_{OS}}{300°}$$

where ΔV_{OS} is the offset correction at room temperature. From this last expression it is seen that a drift of 3.3 $\mu V/°C$ is added for each millivolt of offset nulled. The above results apply directly to any operational amplifier with no emitter degeneration resistors in the first stage. Emitter resistors reduce the portion of offset correction that is developed on the transistors, and the change in drift above is reduced by the ratio of r_e to the total resistance in the emitter circuit.

A more general view of the effects of the offset compensation current results from consideration of the differential current unbalance it creates in the first stage. This result is independent of the presence of emitter resistors. For a bipolar transistor input operational amplifier it can be shown that the drift effect of current unbalance will be[3]

$$\left(\frac{dV_{OS}}{dT}\right)_c \doteq (200\,\mu V/°C)\log\frac{I_{E1}}{I_{E2}}$$

This equation is plotted in Fig. 1.3. On this graph it can be seen that only a 10 percent current unbalance changes voltage drift by 8 $\mu V/°C$.

Similarly, for an FET input operational amplifier it is found that the effect on drift is expressed by[3]

$$\left(\frac{dV_{OS}}{dT}\right)_c \doteq -2.0\,mV/°C\frac{\sqrt{I_{D1}} - \sqrt{I_{D2}}}{\sqrt{I_{DZ}}}$$

where I_{D1} and I_{D2} are near the zero temperature coefficient drain current I_{DZ}. Figure 1.4 displays this result and points out the extreme sensitivity of FET operational amplifier drift to current unbalance. Note that only a 1 percent unbalance results in an 11 $\mu V/°C$ change in drift. This sensitivity coupled with the high offset voltages in FET types can make the offset null and drift interaction critical. While many FET operational amplifiers use other null approaches to avoid this problem, some are severely limited by it. It is again found that removing 1 mV

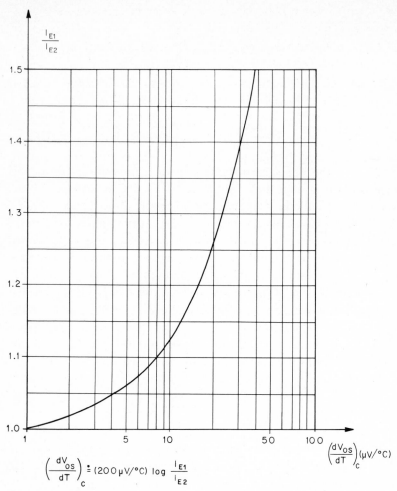

Fig. 1.3 Input offset voltage drift resulting from current unbalance in a bipolar transistor differential stage.

of offset adds around $3\,\mu V/^\circ C$ of drift, but the offsets to be nulled are often much higher with FET input amplifiers.

1.1.3 Null techniques with no drift effect   To avoid the above offset voltage drift created by offset nulling, an offset correction signal with a compensating temperature coefficient can be used, or null can be achieved by adding a dc input signal. With a bipolar transistor input operational amplifier it is fairly simple to supply a temperature-dependent offset correction signal that does not disturb offset voltage drift. This requires a correction current with a temperature coefficient opposite to that of r_e for the amplifier with no emitter degeneration resistors in the above discussion. Since the temperature coefficient of r_e is approximately opposite to that of a forward junction voltage, the required

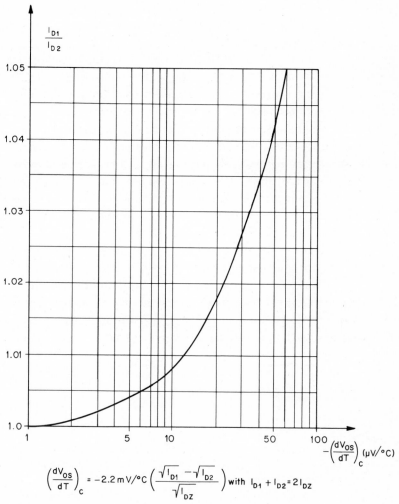

$$\left(\frac{dV_{OS}}{dT}\right)_c = -2.2 \, mV/°C \left(\frac{\sqrt{I_{D1}} - \sqrt{I_{D2}}}{\sqrt{I_{DZ}}}\right) \text{with } I_{D1} + I_{D2} = 2I_{DZ}$$

Fig. 1.4 Input offset voltage drift resulting from current unbalance in an FET differential stage.

compensation becomes fairly straightforward. A correction current derived from a junction voltage, such as V_{BE}, can accurately compensate the thermal variation of r_e, as both characteristics are straight-line functions of temperature. This straight-line dependence is illustrated by the common expressions

$$r_e = \frac{KT}{qI_e} \qquad \frac{dV_{BE}}{dT} = -2 \text{ mV/}^\circ C$$

Both characteristics have temperature coefficients with magnitudes of about $0.3\%/^\circ$ C, and their coefficients are of opposite sign, so that their effects can be made to cancel.

An offset control providing such temperature compensation with a common type of null provision is given in Fig. 1.5. The offset control derives a correction current from the emitter-base voltage of a transistor to develop the appropriate temperature coefficient. Biasing current from R_1 divides between R_2 and the transistor as controlled by their feedback loop. Current in R_2 develops a bias voltage for the transistor, and this voltage in turn controls the portion of the biasing current conducted by the transistor. At equilibrium the current in R_2 will be V_{BE}/R_2. If this current is much greater than the transistor base current, then the offset correction current I_c is also approximately V_{BE}/R_2, as desired. This current is divided by the control potentiometer for offset voltage adjustment. For those amplifiers with input emitter degeneration resistors, the same circuit can be used if resistance is added to the emitter of the compensation transistor. The appropriate resistance value to be added is one which produces a voltage drop equal to that of the degeneration resistors in the amplifier. With this addition, the temperature coefficient of the correction current is lowered to match that of the overall emitter circuit resistance.

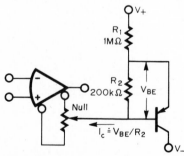

Fig. 1.5 A temperature-compensating offset correction current is derived from emitter-base feedback around a resistor.

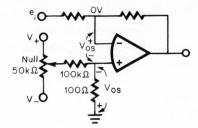

Fig. 1.6 Input offset voltage null in the inverting configuration is provided by biasing the noninverting input.

Alternately, offset voltage null can be achieved by adding a dc correction voltage as an input signal, rather than by using the amplifier null provision. This approach will be illustrated for the basic operational amplifier feedback configurations with resistive feedback, although the same operation can be achieved with other types of feedback. For the inverting configuration, it is simple to null the offset voltage effect with a dc input voltage, as shown in Fig. 1.6. Here the noninverting input is biased away from ground by a voltage that is equal and opposite to the offset voltage. Then the inverting input is at zero voltage level, and no input or feedback current results from V_{OS}.

The analogous null approach for noninverting operational amplifier connections is not as convenient. Figure 1.7 illustrates this null technique for the noninverting amplifier and the unity-gain-follower connections. For the amplifier circuit, a dc correction signal is supplied to the feedback network. Offset correction with the follower is achieved by passing a dc current through a small resistor added in the feedback path. The voltage drop on this resistor is set to cancel the effect of V_{OS}. How-

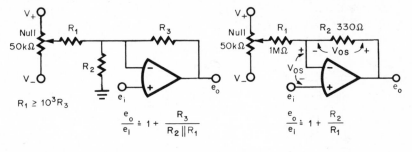

$$\frac{e_o}{e_i} \doteq 1 + \frac{R_3}{R_2 \| R_1}$$

$$\frac{e_o}{e_i} \doteq 1 + \frac{R_2}{R_1}$$

NONINVERTING AMPLIFIER   FOLLOWER

Fig. 1.7 Simple dc signal nulling of offset voltage in noninverting circuits also alters gain.

ever, the null circuits affect the voltage gains of these noninverting circuits, as demonstrated by their accompanying gain equations. While the resultant gain is closely predictable, variations of the null potentiometer have an effect on gain. The gain added to the unity-gain follower by the null circuit produces a departure from the desired gain of unity of the order of 0.03 percent.

These gain errors due to the offset null circuit can be avoided by increasing the shunting impedance presented by the null circuit. Simply increasing resistor levels is unsatisfactory, as this lowers the available compensation current. Instead transistor current sources can be used to supply virtually any current level from a high output resistance. For both polarities of offset null a bipolar current source is required, as illustrated in Fig. 1.8. Here transistors Q_1 and Q_4 are current source-biased transistors having cascode collector biases from the FETs Q_2 and Q_3. Only the difference between the two current source outputs flows in the feedback resistor R_2 to compensate for voltage offset, and this difference is controlled by the null potentiometer. The offset correction is the same for a voltage follower with feedback resistance. With the cascode bias the signal swing is essentially removed from the bipolar transistor collectors and impressed on the FET gate-drain junctions, resulting in much higher output resistance. The residual swing on the collector-base junctions is lower than the original signal by a factor of $r_{ds}g_{fs}$, and the resulting

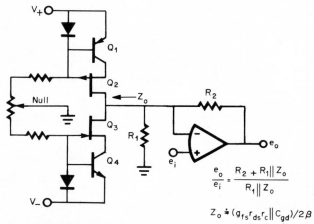

$$\frac{e_o}{e_i} = \frac{R_2 + R_1 \| Z_o}{R_1 \| Z_o}$$

$$Z_o \doteq (g_{fs} r_{ds} r_c \| C_{gd})/2\beta$$

Fig. 1.8 Offset voltage null of noninverting amplifiers without inducing offset drift or gain error is provided by an adjustable, bipolar current source.

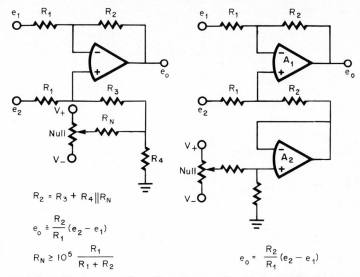

Fig. 1.9 Difference amplifier offset voltage null with the basic null circuit degrades CMR, and so a separate null amplifier may be desirable.

null circuit output impedance is

$$Z_o = (g_{fs}r_{ds}r_c \parallel C_{gd})/2\beta$$

This results in output impedances of the order of $100\,\text{M}\Omega$ and very small-signal shunting and gain error.

Application of a simple dc signal will also remove offset voltage from the difference amplifier circuit, as in Fig. 1.9. Again a dc signal is added to alter the bias at the noninverting input; however, the shunting effect of the control can seriously degrade common-mode rejection (CMR). A very small mismatch between the divider networks connected to the two inputs will significantly lower CMR. To preserve a CMR level near that of the operational amplifier itself, the null resistor R_N must meet the condition indicated. This condition is difficult to meet for the offset control range generally needed. Thus, a fixed resistor offset adjustment is normally made, followed by a CMR adjustment. For variable offset control with high CMR, a follower, shown as A_2, is sometimes added. Low output impedance from the follower avoids CMR disturbance.

Perhaps the most precise means of input offset voltage nulling is an automatic control technique, as shown in Fig. 1.10. In this case, the offset is automatically nulled on command from a periodic control signal.

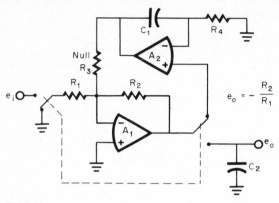

Fig. 1.10 A switched feedback integrator provides automatic offset voltage nulling.

Then any drift or other change in offset voltage will be continually removed by the null circuit. Basically the null circuit is an integrator that is switched into a feedback loop around A_1 when the input is grounded. In this state A_2 integrates any nonzero output from A_1 and supplies an input correction signal to restore a zero output level. The integrator output then holds at this correction level while A_1 returns to the signal-amplifying mode. During the null cycle C_2 holds the most recent output voltage for monitor. An analogous offset-removing operation is performed in chopper stabilization.[3]

1.1.4 Prediction of input current behavior
If the thermal drift nature of operational amplifier input currents is considered, dc performance can often be significantly improved. The most obvious improvement results from the similarity of the two input bias currents. Since they are derived from matched components, the two currents tend to match, and their error effects can be made to cancel. This is commonly achieved by balancing the source resistances seen by the two operational amplifier inputs, as illustrated in Fig. 1.11. In this circuit a resistor is placed in series with the noninverting input, and this equals the parallel combination of the resistors connected to the other input. As a result, the amplifier input currents flow in equal resistances and their effects tend to cancel.

Specifications of the temperature sensitivities of input bias currents and the input offset current are commonly given as an average for a

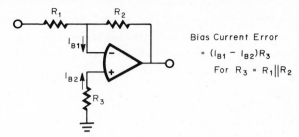

Fig. 1.11 Input bias currents produce canceling error voltages when the resistances of the input current paths are balanced by adding R_3.

specific temperature range.   Generally, performance over the upper end of the temperature range will be better than that specified.   To make use of this better current-drift performance, the nature of the drift should be considered.   Except in amplifiers having added drift compensation, the input current drifts of bipolar transistor operational amplifiers are associated with transistor beta variations.   Since the input bias currents are the base currents of the input transistors, they are inversely proportional to beta.

For silicon transistors the beta variation with temperature produces an input bias current drift approximated by[3]

$$\frac{dI_B}{dT} \doteq CI_B \,(25^{\circ}C) \quad \begin{cases} C = -0.005/^{\circ}C, \ T > 25^{\circ}C \\ C = -0.015/^{\circ}C, \ T < 25^{\circ}C \end{cases}$$

This result is plotted in Fig. 1.12.   From this curve it is obvious that op-

Fig. 1.12 Input bias current drifts of most bipolar transistor operational amplifiers are proportional to beta drift.

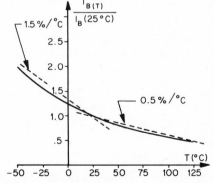

eration in only the higher temperature range would be much less affected by current drift than operation over the entire range. Analogous observations can be made for the input offset current. Since it is the difference between the two input base currents, it is also beta-dependent and is approximated by the curve of Fig. 1.12.

1.1.5 Input guarding

The low input currents and high input resistances of operational amplifiers permit accurate instrumentation of high resistance sources. However, stray leakage paths can greatly increase input currents and decrease input resistance unless the inputs are guarded. Such guarding is generally desirable if stray leakages at the nanoampere level are significant. The guard required is simply a conductive ring surrounding the input terminals and connected to a low impedance point residing at the same dc and signal levels as the inputs. Leakages are absorbed by the low impedance guard ring, and leakage to the inputs is avoided by the equipotential bias of the guard and the inputs. Appropriate guard connections for the inverting and noninverting configurations are shown in Fig. 1.13. For the noninverting amplifier, the feedback resistors must provide a low resistance to drive the guard. However, the resistance seen by the inverting input can be matched to a higher resistance at the other input by adding R_3.

1.2   Preserving AC Performance

Degradation of operational amplifier circuit performance at high speeds can be reduced if the sources of the degradation are identified. Gain error from bandwidth limitations can often be decreased by reducing phase compensation. Reponse speed can be optimized by choosing an

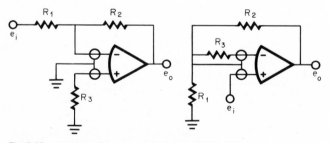

Fig. 1.13 Low impedance guard rings encircling the input terminals and tied to a voltage near that of the inputs absorb stray leakages.

amplifier on the basis of a variety of characteristics. Input impedance loading effects can be accurately anticipated by considering the frequency-dependent nature of the input capacitance. Each of these topics is discussed in this section.

1.2.1   Optimizing gain accuracy   With its high-gain feedback loop an operational amplifier establishes closed-loop characteristics which are primarily determined by the feedback elements themselves. The degree to which these elements fix closed-loop characteristics is determined by the feedback-loop gain. In general, closed-loop gain characteristics deviate from those predicted from the feedback elements by a fractional error ϵ equal to the reciprocal of the loop gain,[3] $\epsilon = 1/A_L$. As a result, gain accuracy is optimized by maximizing loop gain. To permit this a very high open-loop gain is common for an operational amplifier, and heavy feedback or high loop gain is then possible. However, the open-loop gain developed at direct current typically extends only to about 10 Hz with amplifiers having unity-gain phase compensation. Gain accuracy above this frequency is, then, increasingly more difficult to attain, especially when high closed-loop gains further reduce the feedback-loop gain.

However, for the higher closed-loop gain applications, unity-gain phase compensation is not needed, and reduced phase compensation can be used to improve the bandwidth of the open-loop gain. Tailoring the phase compensation to each application improves gain accuracy in this way. The effect of this broadbanding is represented by the Bode diagrams of Fig. 1.14. For closed-loop gains greater than unity the response is tailored as in Fig. 1.14a to ensure stability at a gain level A_1 but not lower. A similar approach increases gain at intermediate frequencies with a modified unity-gain phase compensation, as in Fig. 1.14b. In the latter case, response characteristics are preserved for any gain by controlling the phase shift for a 45° minimum phase margin for all gain levels of unity or above. One further means of increasing gain at high frequencies is the feedforward technique described elsewhere.[3]

To tailor phase compensation for the above results, both theoretical and experimental response evaluations are useful. Response errors are minimized by evaluating loop gain, peaking, overshoot, slewing rate, and settling time, where appropriate. As a basic guide in theoretical evaluation of phase compensation, a phase margin of 45° generally optimizes overall response. With this phase margin a gain peak of 2.5 dB

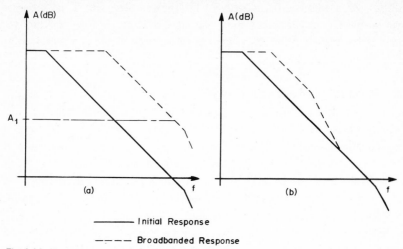

Fig. 1.14   Higher ac gain is attained by modifying phase compensation for (a) closed-loop gains greater than unity and (b) a closed-loop gain of unity.

and a square-wave overshoot of 15 percent without ringing are commonly attained.   For experimental selection of phase compensation the preceding overshoot characteristic serves as a useful guide.   By observing the small-signal square-wave response of the operational amplifier, the appropriate phase compensation can be chosen with generally good results.   Then the other response characteristics are evaluated by testing the closed-loop response for peaking, the open-loop response for gain bandwidth, and the large-signal square-wave response for slewing rate and settling time.   Improvement of any one of these characteristics can then be attempted by monitoring the characteristic while making further adjustments to the phase compensation.   Further details of phase compensation criteria and techniques are presented elsewhere.[3]

Capacitance loading can greatly affect the response characteristics of an operational amplifier and should be included in phase compensation selection.   As illustrated in Fig. 1.15, the load capacitance C_L forms a low-pass filter with the amplifier output resistance.   The resulting response pole increases the slope of the amplifier response, as indicated on the open-loop response curve, and adds phase shift in the feedback loop.   Peaking, ringing, or even oscillation can be produced by this effect at low gain levels unless further phase compensation correction is made. However, the internal phase compensation of many operational ampli-

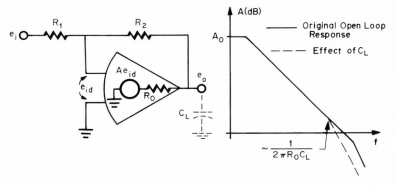

Fig. 1.15 Capacitive loading typically shifts the last response pole back in frequency by its effect upon output resistance.

fiers is not accessible for modification. In these cases additional phase compensation can be added in the amplifier feedback loop, as shown in Fig 1.16. The added components, R_3 and C_f, decouple the amplifier from C_L, as represented by the pole shift and added zero of the compensated Bode plot.

1.2.2 Operational amplifier response characteristics
For high-speed applications a fast operational amplifier is obviously desirable. However, an operational amplifier may be fast in any of five ways. The amplifier appropriate for a given application is chosen by considering the relevant response characteristics, which include bandwidth, full-power response, slewing rate, overload recovery, and settling time. For opera-

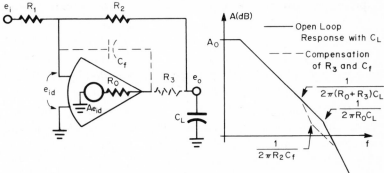

Fig. 1.16 Added phase compensation in the feedback loop modifies the frequency response to restore stability under capacitive loading.

Second, the open-loop response of the operational amplifier should very closely approximate a single-pole response to permit fast settling.[5] This is identified by an accurate, continuous −6 dB per octave response slope. Such a single-pole response is readily modeled by a simple RC response to provide a prediction and theoretical limit of settling time. This yields

$$t_s = \frac{0.367}{f_t} \log 1/\epsilon$$

where ϵ is the fractional error and f_t is the gain-bandwidth product of the feedback loop or the closed-loop bandwidth. To minimize settling time this gain-bandwidth product should be maximized. The maximum gain-bandwidth product attainable with an operational amplifier equals its unity-gain cross-over frequency f_c, and this results for a voltage follower. As the closed-loop gain is increased, loop gain decreases, and f_t decreases from f_c in a manner which can be related by the feedback factor[3] β. The feedback factor represents the fraction of the output which is fed back to the input and defines f_t as

$$f_t = \beta f_c$$

$$\beta = \frac{1}{A_{CLi}} \qquad \text{noninverting}$$

$$\beta = \frac{1}{1 - A_{CLi}} \qquad \text{inverting}$$

where A_{CLi} is the ideal closed-loop gain predicted from the feedback elements for an ideal amplifier.

Equally important to fast operational amplifier response are the characteristics of feedback elements and other circuit conditions. Power supplies must be bypassed close to the amplifier terminals by capacitors having good high-frequency characteristics. Capacitance loading and stray capacitances should be minimized in component selection, component placement, and shielding. To further reduce time constants, resistance levels should be as low as can be driven by the operational amplifier output. Figure 1.19 shows the sources of two of these limiting time constants. First, the stray capacitance C_2 of the resistor must charge and discharge through R_1, and the associated time constant limits settling time. In addition, the charging rate limits both slewing rate and settling time. Shield capacitance C_s produces similar limitations when charged

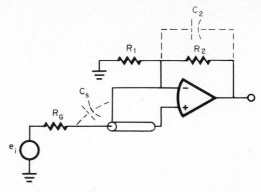

Fig. 1.19 Low feedback resistances and a driven shield reduce the damping effects of component, shield, and stray capacitances.

through high source resistance R_G. However, the effect of C_s can be greatly reduced by driving the shield as shown. In this manner the shield is driven by a signal very nearly equal to the input signal to reduce the signal on C_s by a factor equaling the open-loop gain of the amplifier. However, C_s must now also charge and discharge through the feedback elements.

1.2.3   The significance of input capacitance   The differential input capacitance of an operational amplifier is frequency-dependent rather than a constant. As a result, the capacitance value which should be considered is not immediately apparent. Similarly, the exact effect of input capacitance in operational amplifier circuits is not always obvious. To examine the significance of this capacitance, consider Fig. 1.20, where the capacitance C_I shunts the differential input resistance R_I. Due to this shunting, the amplifier input presents a heavier load to the feedback network, increasing the associated gain error. In addition, the input capacitance introduces phase shift in the feedback loop and degrades the frequency stability and transient response. The degree to which C_I produces these effects can be readily foreseen, but only if the appropriate value of the frequency-dependent C_I is known.

The appropriate value to consider for C_I can be found by examining the frequency-dependence of the capacitance, as depicted in Fig. 1.21. As shown, C_I is a Miller-effect capacitance which is multiplied by the first stage gain A_1 to produce its large, low-frequency value C_{IL}. As the gain

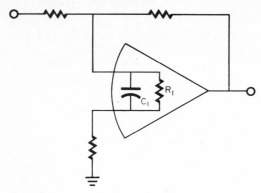

Fig. 1.20 Differential input capacitance shunts the input resistance, producing gain error and feedback phase shift.

of the stage drops with frequency, so does its multiplying effect on collector-base capacitance C_c. The input capacitance then reduces to its small, high-frequency value C_{IH}, which is formed with C_c and emitter-base capacitance C_{eb}. An analogous situation exists for FET input operational amplifiers with the gate-drain and gate-source capacitances.

To compute the effects of C_I the exact expression describing its behavior could be used. Fortunately, however, it is generally possible to approximate C_I with one of its end values C_{IL} or C_{IH}. The transitional

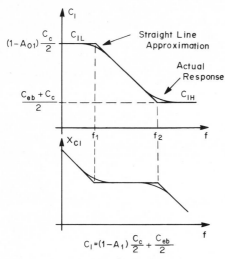

Fig. 1.21 The differential input capacitance varies with frequency over a wide range and has a constant reactance in the transition region.

values of C_I between f_1 and f_2 need not be considered, as the reactance of C_I in this range is constant (see Fig. 1.21). With a constant reactance, C_I appears resistive and cannot present increased shunting or further phase shift. The constant reactance results when the decrease in reactance of C_c is counteracted by the decrease in gain A_1. To decide which end value of C_I to use in calculating its effects, the significance of each can be considered. Typically C_{IL} is much larger than C_{IH}: 500 pF as compared to 3 pF. However, 500 pF does not present a very heavy shunt at low frequencies. This low-frequency value of C_I exists only up to f_1, and this frequency is most often the first open-loop response pole around 10 Hz. Above this frequency C_{IL} produces no additional shunting or phase shift. Thus, if the 500 pF level C_{IL} has no significant effect before its pole at f_1, it may be neglected. Then the 3 pF level C_{IH} will approximate C_I.

To demonstrate the relative significance of each capacitance level, Fig. 1.22 shows their shunting effects on the open-loop differential input

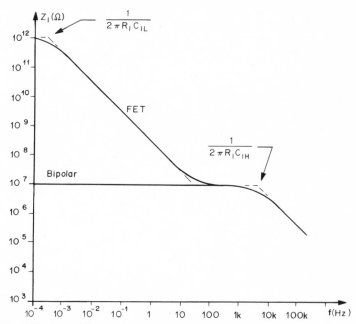

Fig. 1.22 The differential input resistances of FET and bipolar transistor operational amplifiers are far different, but input capacitance shunting equalizes the input inpedances above 10 Hz.

resistance. The input impedance of a typical bipolar transistor operational amplifier is not significantly affected by the large, low-frequency C_{IL}. For most such amplifiers the input capacitance can therefore be approximated by the small C_{IH}. However, the $10^{12}\,\Omega$ open-loop input resistance of an FET input amplifier is heavily shunted by the Miller-effect capacitance C_{IL}. For C_{IL} of 500 pF, the $10^{12}\,\Omega$ input impedance provided by FETs exists only up to 3×10^{-4} Hz. Beyond 10 Hz the FET input impedance is the same as that of the bipolar transistor input. Then for FET input amplifiers, C_{IL} must be considered when high feedback or source impedances rely on the high input impedance to avoid loading error. For a given application the loading error is evaluated considering the closed-loop input impedance, which equals the impedance of Fig. 1.22 multiplied by the loop gain. The falloff of the loop gain increases the rolloff of the closed-loop input impedance.

When it is necessary to avoid the high Miller-effect input capacitance, an operational amplifier having a cascode input stage can be used. Such a stage is presented in Fig. 1.23. This stage preserves the high FET

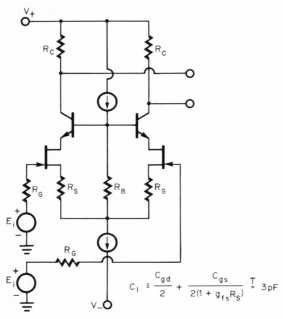

$$C_I \doteq \frac{C_{gd}}{2} + \frac{C_{gs}}{2(1 + g_{fs}R_S)} \doteq 3\,pF$$

Fig. 1.23 Differential cascode biasing removes output voltage swing from the input transistors to reduce differential input capacitance.

input resistance up to about 0.5 Hz by eliminating Miller-effect multiplication of the gate-drain capacitances. The voltage swing on these capacitances is not multiplied by the stage gain and Miller effect is avoided.

1.3 Optimizing Noise Performance

The effects of the input noise voltage and currents of an operational amplifier combine to limit the signal sensitivity of the amplifier. To optimize noise performance, circuit conditions are adjusted for a maximum signal-to-noise ratio. Noise figure is generally a poor parameter for such adjustments on operational amplifier circuits, as will be described.

1.3.1 Signal-to-noise ratio optimization   The noise characteristics of operational amplifiers are commonly modeled by equivalent input noise voltage and current sources, as in Fig. 1.24.[3,6] With this separation of noise sources from the amplifier, the signal sensitivity of the amplifier is immediately apparent regardless of the application. The effects of the equivalent noise sources combine to produce a total equivalent input noise which determines the signal-to-noise ratio of the amplifier. The noise currents i_{ni} flow in the source resistances, creating a noise voltage which combines in a mean-square sense with e_{ni} to produce this total noise.

$$e_{nit} = \sqrt{\overline{e_{ni}^2} + 2\overline{i_{ni}^2}\, R_G^2}$$

When this total noise voltage is minimized, the voltage signal-to-noise

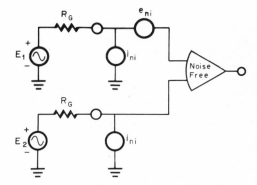

Fig. 1.24 The combined effects of all noise sources in an operational amplifier are represented by equivalent input noise generators.

ratio, and therefore voltage amplifier noise performance, will be optimized. This desired minimum results when source resistance is zero. When the input signal is a current rather than a voltage, an analogous expression can be written which indicates a signal-to-noise ratio maximum for infinite source resistance. However, voltage inputs are more common, and the relative noise performance of different amplifiers can be compared by plotting e_{nit} for each one. This is done in Fig. 1.25 for the basic types of operational amplifiers. From these curves it is seen that bipolar transistor input operational amplifiers commonly provide the lowest noise at low source resistance, while FET input types are best for very high source resistance.

1.3.2 Noise-figure optimization   While noise figure is a more common figure of merit than signal-to-noise ratio for noise performance optimization, it applies in a special class of circuits which does not include most operational amplifier circuits.[3,6] Noise figure is also commonly displayed as a function of source resistance, as in Fig. 1.26. Once again the graphical display indicates the relative merits of different amplifier types at different source resistance levels. However, the curves also suggest that noise performance for a given amplifier is best at that level of R_G which minimizes the noise figure. From this feature comes the common recommendation that source resistance be adjusted to minimize noise figure.

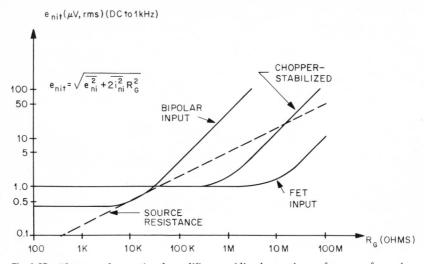

Fig. 1.25   The type of operational amplifier providing best noise performance for a given source resistance can be found using e_{nit} as a figure of merit.

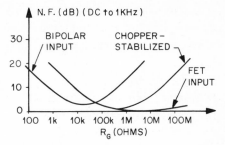

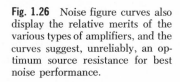

Fig. 1.26   Noise figure curves also display the relative merits of the various types of amplifiers, and the curves suggest, unreliably, an optimum source resistance for best noise performance.

This is in contradiction with the signal-to-noise ratio results above, which indicated that zero source resistance minimized e_{nit} for maximum signal sensitivity. At zero source resistance the noise-figure curves reach infinity, suggesting very poor performance. However, adding source resistance to provide minimum noise figure can only increase the noise voltage developed by input noise currents.

This contradiction commonly results from an overgeneralization of the performance indicated by the noise figure. In general the noise figure does provide a relative comparison between two amplifiers in a given circuit, but it does not necessarily give a reliable comparison between two circuit conditions for the same amplifier. Noise-figure comparison of performance for various source resistances is, then, not always reliable. This practice came from the r-f and audio circuits, in which transformer coupling is common, as represented in Fig. 1.27. When the turns ratio is

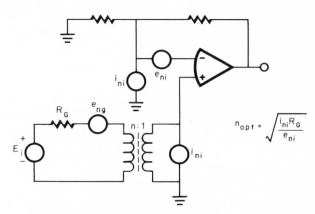

Fig. 1.27   With a transformer-coupled input, a turns ratio variation which increases source resistance, and thereby adds noise, also increases the transformed signal, so that signal-to-noise ratio might be increased.

changed to increase the transformed source resistance, the signal is also increased. This results in maximum signal-to-noise ratio and optimum noise performance at some nonzero value of resistance. In this case the optimum source resistance is the one which minimizes the noise figure, and the noise figure is here a direct measure of noise performance. However, this result of this specific circuit form cannot be applied more generally to operational amplifiers.

1.4   Overload Protection[7]

Voltage and current overloads are common causes of operational amplifier failures. While most operational amplifiers incorporate some overload protection circuitry, few can withstand all the different overload conditions which can damage an amplifier. Some of these overload conditions are obvious, such as input breakdown under excessive input voltages or output overheating under short circuit. Most operational amplifiers are protected to some degree against the latter two conditions. However, other overload conditions which may be peculiar to certain applications are less evident. These include the effects of voltages maintained by capacitors when the power supplies are turned off, and the forward-biasing of integrated-circuit substrate junctions. A voltage retained at an amplifier input by a capacitor or other source can forward-bias and destroy a substrate junction when the negative supply voltage becomes less negative than the input voltage. Overload protection circuits for power supply faults, input overloads, and output overloads are described in this section.

1.4.1   Protection from power-supply faults   The most common power-supply faults in operational amplifier circuits are supply reversals and voltage transients. Damage from these overloads is prevented by the circuits of Fig. 1.28. To protect against damage from voltage reversal, a diode is added in series with each power-supply terminal to block reverse current flow. This protection also prevents forward-bias of an integrated-circuit substrate junction, since a reverse-biased diode will now disconnect the negative supply. However, for the latter protection alone resistors can be added in series with the inputs to limit the substrate current flow to a few milliamperes.

Transient protection is provided by the zener diode clamps and the

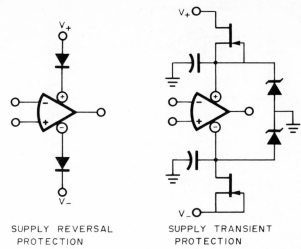

SUPPLY REVERSAL
PROTECTION

SUPPLY TRANSIENT
PROTECTION

Fig. 1.28 Damage from power-supply faults is prevented by using diodes to block current under a voltage reversal and by clamping the supply terminals to remove transients.

voltage-absorbing FET current sources shown. The zener diodes are chosen for ON voltages greater than the normal supply voltages but less than the maximum supply ratings of the amplifier. Under these conditions the zener diodes will be OFF under normal supply voltages, and they will clamp the impressed voltages under supply transients. The current source-connected FETs are chosen for I_{DSS} levels above the normal current drains of the amplifier. Below the I_{DSS} level the FETs are below pinchoff and appear as small resistances in series with the supply lines. Under a transient the zener diodes turn to ON to clamp the supply terminal voltages, and their current drains raise the FET currents to I_{DSS}. Now the FETs are in pinchoff and appear as high impedance current sources to support the excess voltages. As long as the transients do not create voltage breakdown in the FETs, the transient currents are limited to I_{DSS}.

1.4.2 Input protection   Overload conditions at operational amplifier inputs are essentially those of excessive common-mode and differential voltages. Either condition can induce a voltage breakdown that will damage or destroy the input transistors. Because of the precise matching needed between the input transistors, only minor damage from breakdown can significantly degrade dc input characteristics. Such

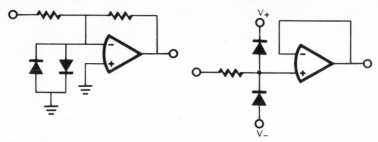

Fig. 1.29   Input clamps protect against any level of input transient that does not force excessive current from the input resistor through the diodes.

damage can result from moderate differential input voltages with bipolar transistors having the typical 6 V emitter-base breakdown voltage. Input stages having FETs are commonly less sensitive to overloads of this magnitude, but they are more sensitive to the low-energy, high-voltage static discharge frequently encountered.

Protection from very high input voltages can be provided by the diode clamps of Fig. 1.29. For both the inverting and noninverting configurations shown, the diodes limit the voltages reaching the amplifiers to safe levels without restricting signal swing. Input transients of thousands of volts can be withstood in this manner, as long as the diode currents are adequately limited by the input resistors. To permit amplifier common-mode swing in the noninverting configuration, the clamp diodes are connected to the power supplies rather than ground. Here diode leakages will add to the input error current.

Input protection against differential signals up to the levels of the power-supply voltages is usually incorporated in operational amplifiers. However, higher-voltage overloads can still damage the amplifiers. Further protection against differential overload can be provided, as in Fig. 1.30, as long as the overloads do not raise either input beyond the supply voltage levels. In cases where this latter condition is also possible, the supply-level clamps of the last figure should be added. Once again diode clamps are used, as in Fig. 1.30. In this case the current-limiting resistance is divided equally between the two inputs so that the error voltage drops produced with the input bias currents will tend to match and cancel. Some error does remain due to the difference in input bias currents. Another error with this clamp circuit can result from the input current it draws under overload. This input current can be a serious error in comparator circuits, where high input resistance in the overload state is desired.   To lower the overload input current the input resistors can be increased, but this also increases the error voltage produced with the input offset current of the amplifier.

The compromises of the above clamp circuit can be avoided with the second protection circuit of Fig. 1.30.[8]   In this case the differential input voltage is limited by a high resistance divider for low current under overload.   In normal operation the large resistors would develop significant error voltages, but they are shunted by low FET resistances.   Specifically, this dc error is significant only at the comparator trip point, where the added voltage would produce an offset.   At the trip point the differential input voltage is zero, leaving the diodes OFF and zero gate bias for the FETs. Under this bias both FETs have a low channel resistance r_{ON}, and this resistance produces only a small error voltage with the input bias current.   When the input signal moves away from the trip point, the gate-source voltage of one or the other of the two FETs increases.   This increases the channel resistance of the FET until it reaches the megohm level r_{ds} at pinchoff.   Then the input current must flow through the resistor paralleling this FET and through the other FET, which remained zero-biased.   This leaves an input divider $R_2/(R_1 + R_2)$ to reduce the input signal.

1.4.3 Output protection   The most common output overloads are excess power dissipation when the output is shorted and output-stage breakdown when shorted to a high voltage.   Most operational amplifiers incorporate current-limiting circuits to control this power dissipation.   External current limits can be added if limits are not included in the amplifier or if a lower level of limiting current is desired to protect a load.   Such a reduced current limit may be needed when the output is shorted to a point above or below ground potential.   The added potential in-

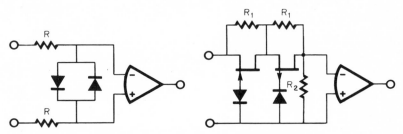

Fig. 1.30 Differential input overloads may be clamped or selectively divided.

creases output-stage dissipation. If short circuit occurs at a voltage beyond the power-supply levels, voltage breakdown can result. Described
below are means for protecting an operational amplifier output from both
excessive output current and voltage breakdown.

One way of providing an external current limit is to connect current
sources in series with the power supplies, as in Fig. 1.31. When the supply current drains are below the design level of the current sources, the
transistors add low resistances in series with the supplies. The bipolar
transistor current sources are then in saturation, adding resistances equal
to r_{SAT} plus 10 Ω, and the FETs are not yet in pinchoff, so they each add
a resistance of r_{ON}. If these small resistances are bypassed as indicated,
they have little effect upon performance. When the supply current
drains reach the design levels of the current sources, the transistors enter
their current source modes with very high output resistances. Only a
small additional current is then needed to develop large voltage drops
across the current sources and to reduce the supply voltage across the
amplifier. By reducing this voltage while limiting current, a power limit
is achieved rather than just a current limit.

A somewhat simpler external current limit is provided by adding a current source in series with the output. The simplicity comes from the
ability of an FET to operate in an inverted mode so that only one FET is

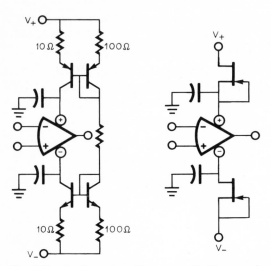

Fig. 1.31 Precise, selectable current limiting can be
added to an operational amplifier by connecting current
sources in series with the power supplies.

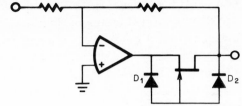

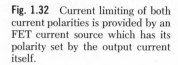

Fig. 1.32 Current limiting of both current polarities is provided by an FET current source which has its polarity set by the output current itself.

necessary, as in Fig. 1.32. For currents flowing into the output terminal, D_2 is reverse-biased, and the voltage on the FET produces a gate-drain leakage current that is conducted by D_1. For this low diode current the voltage on D_1 is too low to forward-bias the gate-source junction, but the diode does connect the gate to the source. This connects the FET as a current source for limiting with the same FET operation as above. Once again the series resistance added by the FET is the low r_{ON} until limiting occurs at I_{DSS}. Since this resistance is inside the amplifier feedback loop, its effect is divided by the feedback-loop gain. When the output current reverses and flows out of the output terminal, D_1 turns OFF and D_2 conducts the leakage current. In this way the gate is connected to the drain for an inverted FET current source. In effect, the polarity of the current source reverses to provide limiting of the opposite polarity of current as well.

If output short-circuit or inductive loads cause the output voltage to exceed one of the power-supply levels, the output stage can be damaged by voltage breakdown. Protection against such overloads is provided by the zener diode clamps of Fig. 1.33. With the zener diodes the operational amplifier output terminal cannot be pulled beyond selectable voltage levels, and the excess voltage is absorbed by R_L. The current-limiting resistor R_L is chosen large enough to protect the zener diodes but not so large as to develop a swing-limiting voltage in normal operation. Since R_L is within the feedback loop, its addition to output impedance is divided by the loop gain. Note that a high voltage on the output terminal

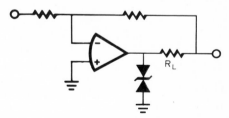

Fig. 1.33 Protection from excessive voltages connected to the output terminal is provided by clamping zener diodes.

also raises the voltage on the inverting amplifier input through the voltage divider formed by the feedback resistors. While this voltage is reduced by the divider, it can sometimes break down the input stage. Where this is possible, the input protection considerations of the last section should be applied.

REFERENCES

1. A. H. Hoffait and R. D. Thornton, Limitations of Transistor DC Amplifiers, *Proc. IEEE*, February 1964.
2. J. G. Graeme, Offset Null Techniques Increase Op Amp Drift, *EDN*, April 1, 1971.
3. G. E. Tobey, J. G. Graeme, and L. P. Huelsman, *Operational Amplifiers: Design and Applications*, McGraw-Hill Book Company, New York, 1971.
4. B. Schultz, Op Amp Follower Has Wide Adjustable Offset Voltage, *Electron. Design*, Oct. 25, 1970.
5. L. Kedson and G. Tempei, Operational Amplifier Frequency Response—It's the Shape that Counts, Parts I and II, *Electron. Design*, July 5 and 19, 1965.
6. J. G. Graeme, Don't Minimize Noise Figure, *Electron. Design*, Jan. 21, 1971.
7. J. G. Graeme, Protect Op Amps from Overloads, *Electron. Design*, May 10, 1973.
8. L. Accardi, Modified 710 Maintains Accuracy at High Input Voltages, *EEE*, October 1970.

2

AMPLIFIERS

While the operational amplifier is by nature a highly versatile component, its utility can be extended in a variety of ways. Described in this chapter are numerous techniques for extending operational amplifier performance in general-purpose applications and in instrumentation amplifier applications. The input impedance presented to a source by operational amplifier circuits can be greatly increased. Output current, voltage, and power can also be increased. Differential outputs can be derived to increase output swing or to accommodate floating loads. Performance in instrumentation amplifier circuits can be improved with guarding and common-mode rejection boosting techniques. Exceptional input-output isolation can be achieved with a number of isolated couplers.

2.1 Increasing Input Impedance

Typical operational amplifiers have input impedances around 10^7 to 10^{12} Ω to avoid loading errors. However, this is not the input impedance which an inverting amplifier circuit presents to the source. In addition, some high impedance sources are significantly loaded by even the common-mode input impedance presented by noninverting circuits. However, the input impedance presented by an operational amplifier circuit can be greatly increased. Generally, the input resistance attainable is limited by stray leakage paths to about 10^{12} Ω. As described in Sec. 1.2.3, this high impedance is shunted at a fairly low frequency by even a 3 pF input capacitance, making the extremely high input impedance difficult to maintain. Nevertheless, that impedance is generally adequate for low source loading, since the source impedance is also limited by stray leakages and capacitances. Input impedance boosting techniques described in this section are those of buffering, bootstrapping, and isolation.

2.1.1 Input buffering   Perhaps the most straightforward way to increase input impedance is with additional buffering. For inverting circuits this is, obviously, achieved by increasing the input resistor, but this adds dc error from input bias currents and requires a similar increase in feedback resistance. By using an FET input operational amplifier, the input error currents are minimized, but the high feedback resistance can pose several problems. In particular, most resistors of extremely high value are less precise, less stable, and have serious bandwidth limitations. Bandwidth is greatly limited by small, internal parasitic capacitances with very high-level resistors. All these difficulties are avoided by the feedback T network of Fig. 2.1. Here low-value resistors produce the same effect as a much higher feedback resistance by their voltage divider action. This network has a short-circuit transfer

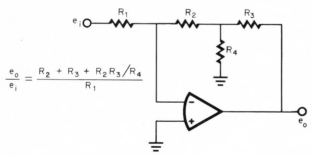

$$\frac{e_o}{e_i} = \frac{R_2 + R_3 + R_2 R_3/R_4}{R_1}$$

Fig. 2.1 The T network adds a voltage divider in the feedback loop for high gain with low-value resistors.

impedance that makes it equivalent to a feedback resistance of[1]

$$R_f = R_2 + R_3 + \frac{R_2 R_3}{R_4}$$

As a result, high input resistance and high gain can be achieved without high-level feedback resistors.

In the noninverting configuration an operational amplifier presents its high common-mode input impedance to the signal source. For an FET input amplifier, this impedance reaches the limitations imposed by stray leakages and capacitances described above. However, this input impedance is lower for a bipolar transistor input amplifier, and the additional buffering from an external FET is sometimes beneficial. Such buffering is provided simply with the circuit of Fig. 2.2. Relatively small offset and gain errors are added by this buffer, although it is essentially a source-follower circuit. Low offset results from biasing the source follower with a matching FET current source to cancel the gate-source voltage shift of Q_1. As shown, the source resistor of Q_1 has a voltage drop set by the current of Q_2 to equal the gate-source voltage of Q_2. Since the two FETs have the same source current level, the gate-source voltage of Q_2 is nearly equal to that of Q_1, and the offset shift is cancelled.

Current source bias also reduces the gain error of the buffer by decreasing the signal-induced current change in Q_1. The residual current change is determined by the net impedance at the output of the buffer, $(r_{ds}/2) \parallel R_{Icm}$, and the current produces an error controlled by the trans-

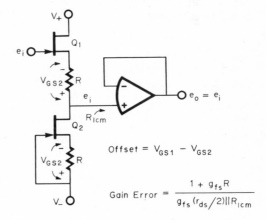

Fig. 2.2 Added buffering from a source follower biased from a matching FET is achieved with low error.

$$\text{Offset} = V_{GS1} - V_{GS2}$$

$$\text{Gain Error} = \frac{1 + g_{fs}R}{g_{fs}(r_{ds}/2)\parallel R_{Icm}}$$

conductance of the follower, $g_{fs}/(1 + g_{fs}R)$.   The resulting gain error, which is generally less than 0.1 percent, is

$$\epsilon = \frac{1 + g_{fs}R}{g_{fs}\,(r_{ds}/2) \parallel R_{lcm}}$$

Additional error is induced by the phase shift of the buffer.   Alternately, the buffer can be connected in series with the inverting input for inverting operation.   In this case, the buffer is enclosed in the feedback loop, so that its gain error is reduced in effect.

2.1.2   Bootstrapping   Controlled positive feedback is sometimes used to increase input impedance by making the output signal supply the input signal current.   As long as the positive feedback does not raise the feedback factor[2] above unity, oscillation will not result.   To bootstrap the input resistance of an inverting amplifier circuit, a second amplifier can be used to supply the input current, as in Fig. 2.3.   In this circuit the inverting amplifier is formed with A_1, and the input current through R_1 is supplied by R_3 rather than by the input signal.   To achieve this the inverting amplifier formed with A_2 derives a signal of $2e_i$ from e_o for driving R_3.   This results in a signal voltage on R_3 which equals e_o so that the circuit input resistance becomes

$$R_I = \frac{R_3 R_1}{R_3 - R_1}$$

If R_1 and R_3 were equal, the input resistance would become infinite, but the attainable resistance is limited by a variety of factors.   The precision of each resistor directly affects the accuracy of the input current cancellation.   If the combined fractional error in resistance values is ϵ, then the input resistance will be no greater than R_1/ϵ.   Should this combined error make the bootstrapped input impedance negative, oscillation can result.   To avoid this, the combined resistance value error should be skewed to assure that the sum of the input and source resistances will always be positive.   Although they are of secondary importance in this stability consideration, the effects of the resistor temperature coefficients and the amplifier gain errors on the bootstrapping accuracy should also be considered.   A further error results from the phase shifts of the operational amplifiers.   As frequency increases, the feedback current becomes shifted in phase from the input signal, and the result is an equivalent input capacitance.

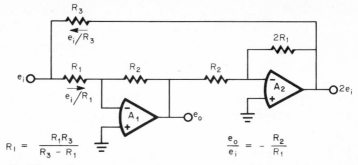

Fig. 2.3 High input resistance for an inverting amplifier is produced with a second amplifier that supplies positive current feedback.

Another bootstrapping technique can be employed to neutralize the input capacitance in noninverting amplifier circuits, as in Fig. 2.4. In this case a bootstrapping current is fed back through neutralizing capacitor C_N to cancel the input current shunted by the common-mode input capacitance C_{Icm}. For noninverting circuits C_{Icm} is the dominant portion of the input capacitance. While the differential input capacitance is also a contributor, it has a very much smaller signal swing and, therefore, a generally negligible effect. The degree to which the feedback current cancels that of C_{Icm} determines the net input capacitance, which is

$$C_I = C_{Icm} - \frac{R_2}{R_1} C_N$$

Somewhat greater feedback can be used to also neutralize the output capacitance of the signal source, as long as the net capacitance at the input remains positive.

The degree of neutralization which can be achieved is limited by the thermal stability of the input capacitance and by the bandwidth and phase shift of the operational amplifier. Thermal variations in C_{Icm}

Fig. 2.4 Positive feedback through C_N neutralizes the input capacitance.

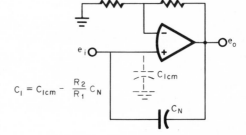

$$C_I = C_{Icm} - \frac{R_2}{R_1} C_N$$

can result in oscillation if the net capacitance of the source and amplifier becomes negative. Due to the large input capacitance variations normally encountered, the neutralizing can only be relied on to provide an order-of-magnitude improvement. This improvement falls off outside the closed-loop bandwidth of the amplifier. Even before the bandwidth limit is reached, the neutralizing is degraded by the phase shift of the amplifier and the resulting phase error of the neutralizing current.

2.1.3 Isolation techniques

Operational amplifier input impedance can also be increased by isolating the input stage from common-mode voltage swing. Without a voltage swing on the input stage, essentially no input current will be developed by this signal. In this way, input impedance can be boosted for common-mode swings that are either intentional or stray signals. The floating power supply of Fig. 2.5 is driven from the amplifier output to remove signal voltage from the internal circuitry of the amplifier. Feedback through the zener diodes maintains fixed bias voltages from the output to the amplifier supply terminals. Then, the only signal on the amplifier internal circuitry is the difference between the input and output voltages. This is the small differential input signal of the amplifier. Then, the input impedance is $Z_I' = AZ_I$, where A and Z_I are the amplifier open-loop gain and open-loop input impedance.

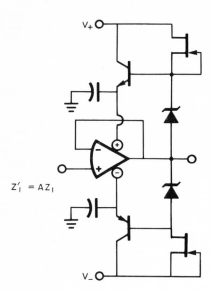

$$Z_I' = AZ_I$$

Fig. 2.5 Zener diodes biased from the output by current sources provide a floating power supply which tracks the signal for high input impedance.

Stray leakage limits Z_i' to around 10^{12} Ω, and the fall in amplifier gain with frequency decreases Z_i'. In effect, this gain decrease results in a significant equivalent input capacitance. However, the common-mode input capacitance normally encountered with a voltage-follower circuit has a contribution to input capacitance that is divided by A.

Additional benefits of this circuit include very high common-mode rejection and extended voltage operation. For a voltage follower the common-mode rejection defines a significant gain error.[2] This error is reduced well over an order of magnitude when the signal swing is removed from the amplifier in this circuit. Signal voltage swings beyond the voltage limitations of the amplifier can be handled, as they merely shift the amplifier and power supply to that level which produces the appropriate output voltage. In this way, the maximum voltage developed across the amplifier is that of the floating supply, and a low-voltage amplifier can be used for a high-voltage swing. Such operation does require high-voltage transistors in the floating supply. Also, the amplifier inputs should be protected from transients by diode clamps to the floating supply, as described in Sec. 1.4. Such transients are developed between the operational amplifier inputs by the response lag of the output to a rapid input signal change.

An isolation technique can also provide high input impedance with single-ended input operational amplifiers, such as most chopper-stabilized and feedforward amplifiers. This approach is particularly useful for such inverting-only amplifiers, since they cannot be connected in the conventional voltage-follower configuration to achieve high circuit input impedance. Instead, these amplifiers can be connected in the voltage-follower configuration of Fig. 2.6. Feedback is now through the signal source, and it forces the input to swing with respect to ground or the output to swing with respect to common. Or, more simply, the feedback forces the amplifier to develop a signal between ground and the isolated power-supply common.

This output signal equals the input signal since feedback maintains near-zero amplifier input voltage. Only this small voltage is impressed on the input stage of the amplifier, unlike the conventional follower, where the entire signal e_i is impressed on the stage. From this reduced signal swing a boosted equivalent input impedance results, with characteristics identical to those outlined for the previous circuit. As an alternative to this circuit, the ground and common may be defined in a way

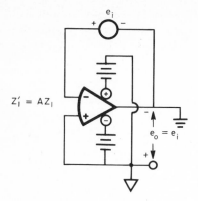

Fig. 2.6 By connecting the signal source as the feedback element, a voltage-follower operation is attained with an inverting amplifier.

opposite to that shown.   This places the power-supply common at the system ground to avoid a separate, floating supply, but this change now floats the signal source.

2.2   Increasing Output Drive Capabilities

To drive most electromechanical devices, such as relays, displays, and speakers, greater output drive than general-purpose operational amplifiers provide is required.   In these applications it is desirable to boost the output current, voltage, or power of the amplifier.   A variety of circuits providing such output boosting are described in this section.   Each of these circuits can be enclosed in the amplifier feedback loop so that the effects of their gain errors and offset voltages are divided by the gain of the operational amplifier.

2.2.1   Output current boosting   The simplest way to boost output current is to add an output emitter follower or source follower.   To efficiently increase both polarities of current, a complementary follower stage is required.   In this case the followers must be biased to set their quiescent currents, and it is generally desirable to protect the transistors with an output current limit.   These elements are incorporated in the complementary emitter-follower circuit of Fig. 2.7.   Class A-B biasing is developed by the diodes, and current limiting is performed by Q_3 and Q_4[2]. With diode biasing the emitter followers Q_1 and Q_2 remain turned ON at the zero crossover, avoiding crossover distortion.   The diodes are biased from an FET current source chosen to supply the maximum base current needed by Q_2.

The booster performance is largely determined by the types of components used. Quiescent current in the emitter followers is determined by the relative junction voltages of the diodes and transistors, as well as by the emitter resistors. Current limiting occurs when the voltage drop on one or the other of the emitter resistors reaches the level needed to forward-bias the associated current-limit transistor. When forward-biased, the current-limit transistor absorbs the base drive current of the corresponding emitter follower. With this limiting the power dissipation in the emitter follower is controlled, and it is the dissipation capability of this transistor which places a maximum on the safe output current. Another limit to the output current is imposed by the base drive current available to the emitter followers from the amplifier and current source. To extend this limit, a Darlington pair or complementary feedback pair can be used in the output stage, as described in Sec. 2.2.3.

Since most operational amplifiers have output current-limiting circuitry, it seems inefficient to add another current-limit circuit in a current booster stage. In some cases this duplication can be avoided by using a booster circuit whose output current is directly controlled by that of the operational amplifier. With a direct control such as this, limiting the amplifier output current will simultaneously limit the booster current. A current booster that operates in this way is shown in Fig. 2.8. Here the added current is supplied by Q_2 and Q_4 as controlled by the amplifier supply currents through Q_1 and Q_3.

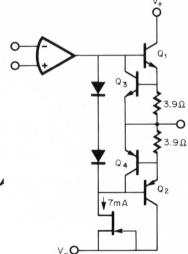

Fig. 2.7 A complementary emitter-follower output current booster with class A-B biasing and current limiting.

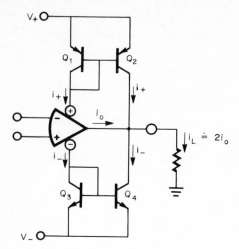

Fig. 2.8   A current booster with controlled current gain is achieved as shown so that the current limit of the operational amplifier will also limit the booster current.

If the like transistors are matched and thermally coupled, Q_1 to Q_2 and Q_3 to Q_4, the currents in the booster transistors will match those in the respective supply current-sensing transistors. When the amplifier output current i_o is zero, the two supply current drains i_+ and i_- are the same, and this current causes an identical current to flow from Q_2 to Q_4. A nonzero output current i_o that increases the current drain from the positive supply creates a matching current increase in Q_2 that is also supplied to the load. Analogously, an output current i_o drawn from the negative supply is matched by an increase in load current from Q_4. Thus, the load current is $i_L \doteq 2i_o$. When the amplifier output current is limited, so is the supply current drain, and the booster current is thereby limited.

The accuracy of the current gain provided by the booster of Fig. 2.8 is primarily determined by the matches of the transistor pairs. Mismatch of a transistor pair will make the associated booster transistor current differ from its biasing supply current. To describe this current-gain error, the junction equation can be used to express the error in terms of emitter-base voltage mismatch. From the junction equation, an emitter-base voltage drop is expressed as[2]

$$V_{BE} = \frac{KT}{q} \ln \frac{I_e}{I_s}$$

$$= 25 \text{ mV} \ln \frac{I_e}{I_s} \qquad \text{at } 25°\text{C}$$

where I_s is the emitter-base reverse saturation current.   Then, the emitter-base voltage difference between two transistors is

$$V_{BE1} - V_{BE2} = 25\,mV\ \ln \frac{I_{e1}}{I_{e2}} + 25\,mV\ \ln \frac{I_{s2}}{I_{s1}}$$

$$= 25\,mV\ \ln \frac{I_{e1}}{I_{e2}} + V_{OS}$$

Note that V_{OS} above is the emitter-base voltage mismatch resulting when emitter currents are equal, and this is the mismatch normally specified for transistor pairs.

A transistor pair with such a voltage mismatch will produce a current mismatch when the emitter-base voltages are forced to be equal, as in the booster of Fig. 2.8.   For this case of $V_{BE1} - V_{BE2} = 0$,

$$\ln \frac{I_{e1}}{I_{e2}} = \frac{V_{OS}}{25\ mV}$$

Assuming small current mismatch, the logarithm above can be expanded to give an expression that relates the current ratio, or current gain, to V_{OS}.

$$\frac{I_{e1} - I_{e2}}{I_{e2}} \doteq \frac{V_{OS}}{25\ mV} \qquad \text{at } 25°\,C$$

Note that a typical $1\,mV$ mismatch corresponds to only a 4 percent current mismatch.

Booster current gains greater than unity can be achieved by using the technique of Fig. 2.8 above with two different modifications.   First, small geometry diodes can be used for biasing large geometry booster transistors to develop a current gain that equals the ratio of the emitter and diode junction areas.   However, this sacrifices the gain accuracy provided by matched transistor pairs, especially if diode-transistor thermal coupling is poor, and the quiescent current is also multiplied by the booster current gain.   To avoid the higher quiescent currents the diodes can be replaced by resistors which are small enough to keep the booster transistor biased OFF under quiescent conditions.   In this case the current gain is very high, and emitter degeneration resistors must be added to the booster transistors to provide a gain that permits booster current limiting with the amplifier current limiting.

Another simple current booster results from the complementary

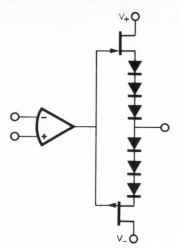

Fig. 2.9 Simple biasing and integral current limiting are achieved with a complementary source-follower current booster.

source-follower approach, as in Fig. 2.9. In this case biasing is simple, separate current limiting is not necessary, and output current is not limited by drive current available from the amplifier. Gate-source voltage bias for quiescent current setting is established by diodes in series with the sources. The number of diodes required depends upon the pinchoff voltages of the FETs and the quiescent current level desired. Since the maximum current through an FET is I_{DSS}, an internal current limit protects the FET without the need for added circuitry. When the current in one of the FETs reaches I_{DSS}, its gate junction forward-biases, and the output current of the amplifier flows through this junction, increasing the load current. For this reason the operational amplifier should also have an internal current limit. While the ease of current limiting and biasing makes this circuit rather simple, the limit and bias are less precise. Wide ranges of I_{DSS} and pinchoff voltage make the prediction of booster performance inaccurate.

2.2.2 Output voltage boosting   Higher output voltage swing can be attained with a voltage follower using the isolation technique of the first circuit of Sec. 2.1.3. For other amplifier configurations the output voltage of an operational amplifier can be amplified to increase swing. This involves a gain stage followed by a buffer stage and requires a separate low-voltage supply for the operational amplifier. These requirements are fulfilled by the circuit of Fig. 2.10. In this circuit Q_5 and Q_6 form a complementary gain stage, followed by a buffer similar to one in the last section. To limit the added gain and ease phase compensation, feed-back is provided through the resistors labeled R_2. The gain of the booster is then $A \doteq R_2/R_1 = 10$, which amplifies the $\pm 10\,\mathrm{V}$ amplifier output to $\pm 100\,\mathrm{V}$. Note that this added stage introduces another phase inversion, so that negative feedback elements should now be returned to the noninverting input of the operational amplifier. Input protection, as described in Sec. 1.4, should be used to protect the inputs from dam-age by normal summing junction signals.

High-voltage common-mode input swings cannot be achieved with this circuit. The common-mode swing continues to be limited by the low-voltage zener supply of the operational amplifier. This is suitable for noninverting gains of 10 or greater, as well as any level of inverting gain. For unity-gain noninverting requirements, the circuit of Fig. 2.5 is a better choice.

When a high-voltage output of only one polarity is required, comple-

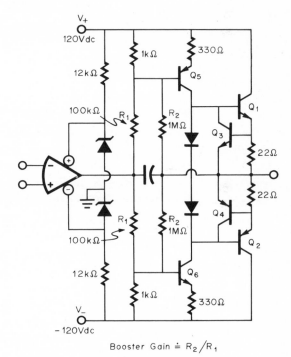

Booster Gain $\doteq R_2/R_1$

Fig. 2.10 High output voltage swing using a low-voltage operational amplifier is achieved with an added gain stage and a derived low-voltage supply.

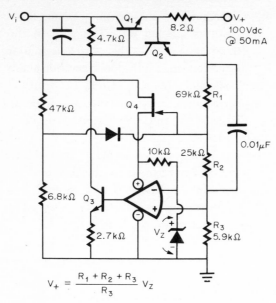

$$V_+ = \frac{R_1 + R_2 + R_3}{R_3} V_z$$

Fig. 2.11 A high-voltage regulator can be built with a low-voltage operational amplifier if a gain stage and low-voltage supply are added.

mentary circuitry is not needed, and the previous circuit simplifies. A common requirement of this type is a high-voltage regulator, which can be realized with the circuit of Fig. 2.11. This circuit parallels the previous one, with gain stage Q_3, emitter follower Q_1 and current limiter Q_2. A feedback voltage divider to one amplifier input establishes a comparison with the reference zener diode. For the low-voltage amplifier a supply voltage is derived through Q_4 from the output itself. Bias from the output avoids the ripple present on the unregulated input voltage and provides a stable bias to the reference zener diode. As a result, the ripple rejection of the regulator is significantly improved. Such output-dependent bias does, however, make zero output voltage a stable state, since the reference is then zero also. This is overcome with the turn-on diode biased from a voltage divider as shown. Load regulation is outstanding because of the high gain of the operational amplifier, but lower gain might be desirable when transient response is considered. By trading gain for bandwidth, a more rapid transient recovery can be achieved. Output accuracy and drift are essentially determined by the feedback voltage divider and the reference zener diode.

2.2.3 Output power boosting
Increased output power is, of course, provided by the output current and voltage boosting circuits described above. However, for higher power requirements, more extensive current or voltage boosting is needed. A circuit which develops high current gain for power boosting is that of Fig. 2.12. It will supply an 8 Ω load with a peak power of 20 W. To supply high currents the power booster has three complementary stages of current gain, with class A-B biasing and current limiting for the output stage. The first stage, consisting of source followers Q_1 and Q_2, buffers the amplifier from the booster and supplies current to the bias regulator Q_3.

Class A-B bias of the output is established by Q_3, which is a simple shunt regulator. With Q_3 shunted by R_1, the collector-emitter voltage of the transistor will equal the voltage on R_1 plus the emitter-base voltage V_{BE3}. For the biasing shown, the voltage on R_1 is set primarily by the current from R_2, as the base current of Q_3 is negligible. The current in R_2 is set by the feedback of Q_3 at V_{BE3}/R_2, and this establishes a voltage on R_1 that is also a multiple of V_{BE3}. Since any given multiple of V_{BE3} can be developed, the output-stage quiescent current can be adjusted

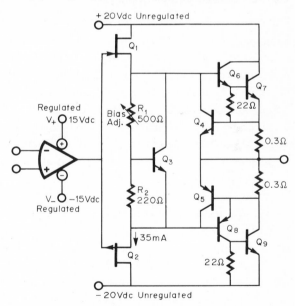

Fig. 2.12 Multiple current gain stages with class A-B output biasing and current limiting provide high power for low impedance loads.

easily. For good bias stability Q_3 should be mounted on the heat sink with the output transistors where thermal feedback will adjust V_{BE3}.

The output stages and the booster power supply are chosen for power efficiency. Compound output transistor structures provide high current gain with low quiescent current drain. On the positive side, the Darlington structure assures high base drive current to the output transistor, while the quiescent drain of the drive transistor is low. A similar condition holds for the complementary feedback pair on the negative side, except that the output power transistor is in a common-emitter configuration so that it can be an n-p-n for lower cost. Current limiting is provided by the feedback clamping of Q_4 and Q_5, as in previous circuits. Power is supplied to the booster from the unregulated output of the amplifier power supply. This avoids the need for voltage regulation at high current levels and greatly simplifies the power supply. Only minor output ripple results from this simplified biasing, as the complementary form of the booster provides high ripple rejection.

2.3 Differential Output Amplifiers[3]

Direct drive of electromechanical devices such as solenoids and indicators frequently requires a greater voltage swing than is available with most operational amplifiers and common supply voltages. At best, a 25 V p-p swing can be obtained from most general-purpose operational amplifiers on ± 15 V dc power supplies. In this case only 5 V of the 30 V total supply voltage is lost for biasing. For greater output voltage swings it is common to resort to more specialized high-voltage amplifiers and power supplies. The cost of these special-purpose amplifiers and a separate power supply is several times that of similar lower-voltage operational amplifier circuits.

Frequently this compromise can be avoided by making further use of the differential nature of operational amplifiers. In particular, by using a differential output operational amplifier, as in Fig. 2.13, voltage swings far greater than the total supply voltage can be achieved. As shown, a 50 V p-p swing can be provided by a differential output amplifier on only the 30 V available from common ± 15V dc supplies. This is a result of opposite polarity outputs. The outputs develop opposing 25 V p-p swings, so that the net swing developed between the two outputs is 50 V p-p. This output can be applied to any floating load not requiring ground

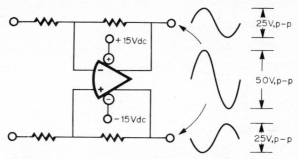

Fig. 2.13 The two outputs of a differential output operational amplifier have opposing swings, and this results in an overall output swing which is twice that of single-ended output amplifiers.

reference. To further improve cost, this differential output performance can be approximated with two single-ended output amplifiers. One such circuit is that of Fig. 2.14. In this case a second, inverting amplifier is added to provide the opposite polarity output. This added output, however, has additional phase shift introduced by the second amplifier. Phase-shift error limits the accuracy of this circuit to a much lower bandwidth than commonly imposed by amplitude error.

A second differential output amplifier formed by two single-ended output operational amplifiers is illustrated in Fig. 2.15. In this circuit the phase shifts to the two outputs are the same and greater gain-accuracy bandwidth is attained. Input impedance is also higher than in the last circuit, since the signal drives amplifier inputs rather than a summing resistor. A differential signal input is also formed, in this circuit, to reject common-mode signals. The common-mode rejection for the circuit equals the average of those of the two operational amplifiers. While a differential signal input could, ideally, monitor a floating source, some

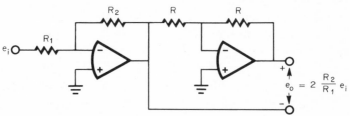

Fig. 2.14 An added inverter develops an opposing output for differential output operation.

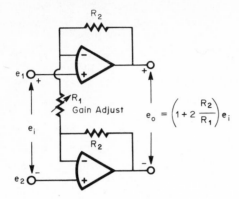

$$e_o = \left(1 + 2\,\frac{R_2}{R_1}\right)e_i$$

Fig. 2.15 Parallel connected non-inverting amplifiers provide a differential output and a differential signal input.

ground return resistance must be provided to conduct the input bias currents of the amplifiers.

To develop a differential output the circuit shown uses a common feedback current. This current is established in R_1 by feedback which forces the differential input voltages of both amplifiers nearly to zero. Then the voltage across R_1 equals e_i, and this defines the feedback current. The feedback current flows out of the feedback resistor of one amplifier, through R_1, and into the feedback resistor of the second amplifier. Because the currents in the two feedback resistors flow in opposite directions, the outputs are of opposite polarity. A swing limitation can result at low gain as a result of the common-mode swing limitation of the operational amplifiers. At unity-gain the high output voltage swing requires a correspondingly high input swing that would exceed the common-mode capability of either amplifier alone. To permit this swing condition the input signal would have to swing differentially about ground so that the signal would be divided between the two amplifiers. If, instead, one input terminal is grounded, full output swing can be developed only for gains of 2 or greater.

A third differential output circuit is formed by connecting two differential amplifier circuits in parallel, as in Fig. 2.16. Opposing output signals result from reversing the input connections on the two differential amplifiers. Once again, a differential signal input is produced along with the differential output. Although the input impedance is low in this case, the inputs will monitor a floating source without additional input shunting to ground. Common-mode rejection is determined by both the operational amplifiers and the feedback networks. For high common-

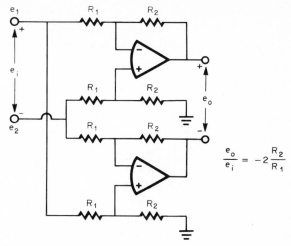

$$\frac{e_o}{e_i} = -2\frac{R_2}{R_1}$$

Fig. 2.16   Parallel connected differential amplifiers result in a differential input, differential output amplifier using single-ended output operational amplifiers.

mode rejection the feedback networks must be experimentally matched by trimming.

2.4   Instrumentation Amplifiers

Voltage amplification requirements in instrumentation are often served by differential amplifiers, commonly called instrumentation amplifiers. Basically, these amplifiers have a differential input and a feedback commited for voltage gain.   While most operational amplifiers do have differential inputs, feedback is generally applied to one of these inputs, leaving only one signal input.   When the feedback network is matched by an identical network to the second input of the operational amplifier, a differential signal input is formed, as in the elementary differential amplifier configuration.[2] Described in this section are simplified instrumentation amplifiers, some techniques for input guarding, methods of boosting common-mode rejection, and a linear gain control.

2.4.1   Simplified instrumentation amplifiers   In the elementary differential amplifier formed with an operational amplifier, voltage feedback to the input results in a relatively low input resistance.   The input resistance is only that of the two input resistors, and significant loading error

can be introduced by this lower input resistance. To retain high input resistance, the conventional voltage feedback can be replaced by current feedback as in Fig. 2.17. In this case the input is buffered from the shunting of the feedback resistor R_F by the β of the input transistors. This gives a differential input resistance of

$$R_I \doteq 2\beta \, (R_E \parallel R_F) \qquad \text{if } R_E \gg r_e$$

where r_e is the dynamic emitter resistance. Common-mode input resistance is somewhat higher, since common-mode signals are impressed on the feedback resistors and the current source more than on the emitter resistors. As a result, common-mode input resistance is primarily determined by the two feedback resistors labeled R_F, and will be

$$R_{Icm} \doteq \beta R_F/2$$

Controlled gain is also provided by the feedback, as it supplies a current proportional to the input signal e_i. When an input signal is applied, it is approximately transferred to the emitter resistors by the emitter-follower action of the input transistors. Initially this produces a current unbalance in the two transistors and, thereby, an unbalance in the emitter-base voltages. The voltage unbalance results in an error in the volt-

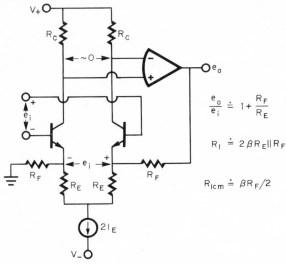

$$\frac{e_o}{e_i} \doteq 1 + \frac{R_F}{R_E}$$

$$R_I \doteq 2\,\beta R_E \| R_F$$

$$R_{Icm} \doteq \beta R_F/2$$

Fig. 2.17 Current feedback to the emitters of a differential stage retains high input impedance for an instrumentation amplifier.

age transferred to the emitter resistors. However, the current unbalance also creates a voltage between the operational amplifier inputs to generate a feedback correction signal. The operational amplifier returns its input voltage to zero by supplying a feedback current to one emitter in order to equalize the transistor currents. This feedback current flows in a resistor R_F, and the output voltage required to supply the feedback current is

$$e_o = \left(1 + \frac{R_F}{R_E}\right) e_i$$

A matching resistor R_F is connected to the opposite emitter to balance the signal currents under common-mode swing and eliminate the common-mode gain.

While this instrumentation amplifier configuration is rather simple, it can have significant gain linearity and common-mode rejection errors. The gain linearity error is primarily a result of transistor mismatch. Unless the dynamic emitter resistances of the two transistors track over wide ranges of emitter current, the feedback-corrected signal established on the emitter resistors will not always equal the input signal. Then, a signal-dependent gain error is introduced which is not corrected by the feedback action described above. Any mismatch in the resistors labeled R_F will result in unequal signal currents under common-mode swing. This current difference is supplied by the operational amplifier to maintain equal collector currents and zero voltage at its own inputs. To supply this current the output must produce a signal voltage that is a common-mode rejection error. When the gain of this circuit is varied, both R_F resistors are changed, and they must be very accurately rematched to maintain high common-mode rejection.

Significantly improved linearity and simplified gain adjustment are provided by more complex instrumentation amplifiers, such as the three-operational amplifier type shown below (Fig. 2.21). Another instrumentation amplifier circuit achieves these performance improvements with only two operational amplifiers if either higher gains or limited common-mode swing are acceptable. This is the circuit of Fig. 2.18, which is basically two interconnected amplifier circuits. Highly linear gain is provided by the operational amplifiers with their feedback. A single resistor R_G adjusts the gain of this circuit without introducing common-mode errors.

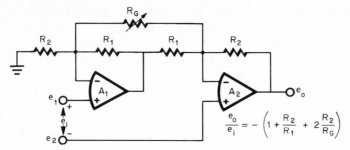

Fig. 2.18 An instrumentation amplifier formed with the two amplifier circuits shown has a single resistor gain control and is well suited for gains of 10 or greater.

To visualize the operation of the circuit of Fig. 2.18 it is beneficial to first neglect R_G and to consider the signal at each input separately, using superposition. Under these conditions the signal e_2 alone is amplified by a simple noninverting amplifier formed with A_2, and the output signal is $(1 + R_2/R_1)e_2$. From the signal e_1 the noninverting amplifier formed with A_1 will develop an output signal of $(1 + R_1/R_2)e_1$. This output signal is presented to A_2 as an input to an inverting amplifier having a gain of $-R_2/R_1$, and results in an output of $-(1 + R_2/R_1)e_1$. Using superposition, the preceding outputs from the two input signals are combined, and this results in an output due to the input difference signal alone of $-(1 + R_2/R_1)e_i$. Including the gain-control resistor in the circuit adds a summing current of e_i/R_G to each summing junction and modifies the gain equation to

$$\frac{e_o}{e_i} = -\left(1 + \frac{R_2}{R_1} + 2\,\frac{R_2}{R_G} \right)$$

The performance of this two-amplifier circuit is limited by the normal operational amplifier circuit errors, resistor matching, and a common-mode swing characteristic. In this case the common-mode rejection errors of the two amplifiers tend to cancel, but the other amplifier errors are additive. From resistor mismatch a nonzero common-mode gain is developed that equals the fractional mismatch. The combined differential and common-mode input swing is limited by A_1. Since A_1 amplifies this combined signal, its full output swing capability is not available for the differential signal alone. For circuit gains of 10 or more the differential signal would not be likely to exceed one-tenth of the swing range, so that essentially a full common-mode swing could be handled. However, for lower gains more of the output swing range of A_1 may be required for the differential signals, and the common-mode swing range must then be restricted.

2.4.2 Input guarding   In applications requiring very high common-mode rejections (CMR), small parasitic capacitances at the inputs can greatly degrade performance.   To illustrate the effect of such capacitance, the common-mode and differential components of the input signal are modeled separately in Fig. 2.19.   The common-mode input signal e_{icm} is attenuated somewhat by the networks formed with the source resistances and the parasitic capacitances.   These capacitances are composed of the amplifier input capacitances and stray capacitances, such as those of cable shields.   Unless $R_{G1}C_{I1} = R_{G2}C_{12}$, the attenuation of e_{icm} at the two inputs will be unequal, resulting in a differential input error signal.   The error signal is amplified along with the desired signal e_i and introduces a common-mode rejection error.   For high CMR the allowable mismatch in attenuation is extremely small.   A CMR of 120 dB, or a common-mode rejection ratio (CMRR) of 1 million to 1, allows an overall mismatch of only 1 part per million when the differential gain is unity.   Such a match is not feasible at frequencies where capacitance shunting is significant.

Input guarding resolves the above problem for stray capacitances, like shield capacitances, at the inputs.   The effect of the amplifier input capacitances is removed, using the techniques of Sec. 2.4.3.   For input guarding the cable shields are not grounded but are driven by a signal equal to the common-mode signal.   Then, the only signal on the shield capacitance is the differential input signal, and e_{icm} passes unattenuated to both amplifier inputs.   However, in practice a shield drive signal equal

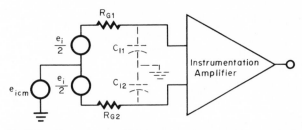

Fig. 2.19 Unequal parasitic input capacitances cause unequal attenuation of common-mode signals at the two instrumentation amplifier inputs, resulting in degraded common-mode rejection.

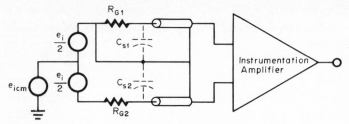

Fig. 2.20 When $e_{icm} \gg e_i$, input guarding to avoid CMR errors from shield capacitances can be provided by driving the shield from one of the inputs.

to e_{icm} is not directly present as modeled in Fig. 2.19. This signal is the average of the two input signals and must be derived. For a common-mode signal which is large compared to the differential input signal, $e_{icm} \gg e_i$, the shield can be driven directly from one of the inputs, as shown in Fig. 2.20. This results in almost ideal input guarding, since the signal at either input is then essentially equal to e_{icm}. It is in this case where $e_{icm} \gg e_i$ that input guarding is most needed, as the common-mode error signals can equal a large percentage of e_i.

When e_{icm} is not large compared to e_i, the input guard signal must be derived as an average of the two input signals. This signal is present in the circuit of Fig. 2.17 at the junction of the emitter resistors. A method of achieving this for a common differential amplifier configuration[2] is illustrated in Fig. 2.21. Here the outputs of the first two amplifiers are summed on a simple resistive divider. If the two resistors are identi-

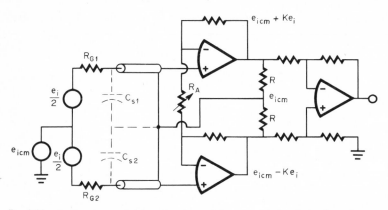

Fig. 2.21 An input guard drive signal equal to e_{icm} can be derived as shown to eliminate CMR error from shield capacitances.

cal and the two amplifier gains are equal, the signal at the center of the divider will be the desired e_{icm}. With this signal driving the cable shields no common-mode swing will be developed across the shield capacitances, and the unequal attenuation error described earlier is avoided. To drive large shield capacitances at high frequencies, the divider resistance should be low, or a buffer amplifier should be added.

2.4.3 Improving common-mode rejection   With stray capacitance effects removed by the input guarding above, the major remaining sources of CMR error are within the instrumentation amplifier. Common-mode swing on the amplifier results in error signals from circuit unbalances and from the common-mode gain of the amplifier.[2] Input capacitance unbalance at the two inputs results in unequal input attenuation of the common-mode signal e_{icm} as described above. Similarly, other unbalances between the two sides of a differential amplifier mismatch the two gains presented to a common-mode signal and create a differential error signal. Once again, at unity-gain, a CMR of 120 dB, or a CMRR of 1 million to 1, requires that unbalances not result in an error of more than 1 part per million referred to the inputs. Such accuracy is not practical over a range of frequencies using the straightforward instrumentation amplifier approaches.

Rather than attempt to control the instrumentation amplifier error sources with such high precision, the common-mode swing can be removed from the amplifier. In this way the circuit unbalances and common-mode gains are circumvented. One method of achieving this is to float the power supplies of the input circuitry, as in Fig. 2.22. The basic circuit is that of a common differential amplifier[2] formed by A_1, A_2 and A_3, and the approach parallels the input guarding technique of Fig. 2.21. As before, a signal equal to the common-mode swing is derived from the input amplifiers and is used to drive cable shields for input guarding. In this case the derived e_{icm} also drives the common of the floating input power supply through buffer A_4. As a result, all bias potentials in the input amplifiers track e_{icm}, and no common-mode swing remains across any circuitry in these amplifiers.

Instead the signal e_{icm} is across the operational amplifier buffer A_4. The finite CMR of this operational amplifier results in an output error, so that the input common-mode swing is not totally removed. However, the net CMR of the input amplifiers is now increased by that of the buf-

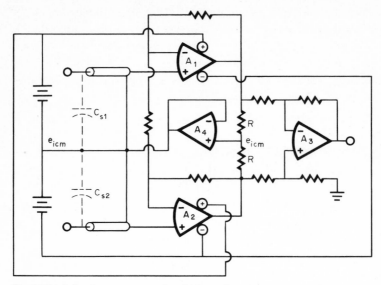

Fig. 2.22 A floating power supply which tracks e_{icm} removes common-mode swing from the input amplifiers to boost CMR.

fer amplifier. The overall CMR of the instrumentation amplifier includes that of the output difference amplifier A_3. The entire common-mode swing is present at the two inputs of this difference amplifier. This swing is rejected to a degree determined by the CMR of the operational amplifier A_3 and by the matching of the associated resistors. While the CMR of this output amplifier is limited by its circuit unbalances, the effect of its CMR on the overall instrumentation amplifier is reduced by the gain preceding it in the input. Overall CMRR is expressed by

$$\text{CMRR} = \frac{\text{CMRR}_{1,2}\text{CMRR}_4}{1 + \text{CMRR}_{1,2}\text{CMRR}_4/(A_{\text{CL}_{1,2}}\text{CMRR}_3)}$$

where the above CMRRs are those of the operational amplifiers with corresponding numbering except for CMRR_3. This last CMRR is that of the overall difference amplifier formed with A_3. The gain $A_{\text{CL}_{1,2}}$ above is the closed-loop gain of the input amplifiers. Typically the CMRR is limited to 1 million to 1 by stray leakage paths.

A second means of increasing CMR is to remove the common-mode swing from the input signal before it reaches the instrumentation amplifier.[4] This is achieved with a common-mode feedback loop, as in Fig. 2.23.   The common-mode swing in the instrumentation amplifier is sensed as before in Fig. 2.21.   By connecting this sense point to one input of an operational amplifier as shown, the signal at the sense point is compared to zero.   Any nonzero common-mode swing in the instrumentation amplifier creates a large signal at the output of the high-gain operational amplifier.   This output signal supplies a canceling common-mode input swing to the instrumentation amplifier inputs to restore a near-zero common-mode input.   In this way the common-mode swing on the amplifier is reduced by the open-loop gain of the operational amplifier, and CMR is increased by the same factor.

This technique alters other characteristics as well.   Both common-mode and differential input signals are attenuated by the input voltage dividers.   As a result, the sensitivity of differential signals to instrumentation amplifier drift is increased.   On the other hand, the common-mode voltage range is increased by the dividers.   The common-mode slewing rate can also be extended, since the instrumentation amplifier does not have to slew with this signal.   Instead the common-mode slewing rate is determined by the operational amplifier.   Differential input resistance is also affected by the dividers and is decreased to $R_i' = 2(R_1 + R_2)$.

2.4.4   A linear gain control   Ideally, the gain of an instrumentation amplifier would be linearly adjustable with a potentiometer.   This would

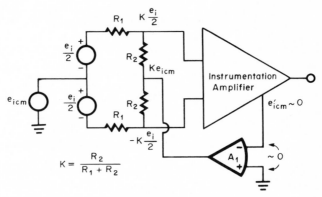

Fig. 2.23   By means of a feedback loop, the common-mode swing is removed from the input signal to increase CMR.

provide a direct indication of gain setting from a potentiometer setting. However, the gains of common instrumentation amplifiers do not bear a linear relationship with any one resistor. For the common circuit modified in Fig. 2.21 the gain is expressed by[2]

$$A = 1 + \frac{K}{R_G}$$

where K is a constant and R_G is the gain-controlling resistor. This nonlinear relationship can be replaced by further modifying the circuit as in Fig. 2.24. In this case the input amplifiers are at unity-gain, and the circuit gain is established in the output amplifier by means of A_4. This added amplifier acts as an attenuator in the feedback loop to vary the relationship between the output voltage and the feedback voltage. At the output of A_4 the feedback voltage must equal e_i, as in the basic differential amplifier connection. Then the output voltage will be

$$e_o = -\frac{R_G}{R_1} e_i$$

Thus, gain is linearly related to the control potentiometer R_G.

Other performance characteristics are also affected by this circuit modification. The output resistance of A_4 will unbalance the resistor networks around A_3 and degrade common-mode rejection somewhat. At low frequencies this output resistance is minimized by feedback, but it can be significant in higher-frequency applications. Voltage offset and

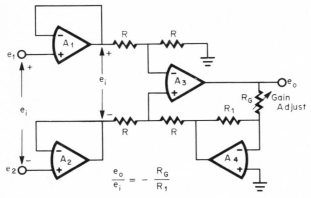

Fig. 2.24 By adding feedback attenuator A_4, a linear gain control is achieved.

drift errors are also increased with this circuit modification. Since the input amplifiers are now at unity-gain, the input signal is not amplified before it is added to the offset voltage of A_3. The net equivalent input offset voltage will be

$$V_{OS} = V_{OS1} - V_{OS2} + V_{OS3} - \frac{R_1}{R_2} V_{OS4}$$

2.5   Isolation Amplifiers

Amplifiers having very high input-output ground isolation can be formed with operational amplifiers and isolated couplers. Such isolation amplifiers exhibit common-mode voltage ranges of thousands of volts, isolation resistances of $10^{11}\ \Omega$, and common-mode rejections of 120 dB. With an isolation amplifier, signals which are far removed from the ground of the monitoring system are accurately measured, and medical instrumentation is made far safer by the high impedance isolation from ground. For internal medical monitoring, such as with cardiac catheters, a mere 20 μA current created by a ground line voltage drop can be fatal, and so ground isolation is highly desirable.

The desired isolation characteristics can be achieved with a variety of signal couplers, but they all have nonlinearities when used alone. For more precise instrumentation the effect of coupler nonlinearity can be overcome with linearizing feedback or can be avoided with modulated carrier systems. Several such isolation amplifier circuits are described below.

2.5.1   Feedback linearized configurations   To correct for the nonlinearity of a signal transfer path, feedback can be applied through a signal path having a matching nonlinearity. In this way the nonlinearity is removed within the degree of match of the transfer and feedback paths. This permits accurate, isolated signal transfer with transformers, LED-phototransistor couplers, photoresistors, magnetoresistors, Hall-effect devices, and thermal couplers. Fundamentally similar isolation amplifier circuits can be used with any of these signal couplers, so that the circuits below are illustrated with the economical LED-phototransistor coupler. Both voltage and current isolation amplifiers are described.

Optical coupling through light-emitting diodes (LEDs) and phototransistors provides signal transmission with excellent isolation qualities. Isolation impedances of $10^{11}\ \Omega$ shunted by $1\,\mathrm{pF}$ are readily achieved. However, signal transmission by this optical coupling typically suffers from limited bandwidth, as well as high distortion. Both of these drawbacks can be greatly reduced using operational amplifiers as described below.

The response speeds attained with phototransistors are commonly far lower than those achieved with conventional transistors. While this might first suggest much lower gain-bandwidth products for phototransistors, the primary limitation is instead largely a result of biasing conditions. An operational amplifier can remove the response limit imposed by these bias conditions.[5] Whether biased as an emitter follower or a common-emitter amplifier as shown in Fig. 2.25a, the speed limitation is the same. In both cases the output voltage swing produces an essentially equal voltage swing on the collector-base capacitance C_c. This produces a feedback current to the base i_f described by

$$i_f = j\omega C_c e_o \qquad \text{for } R_L \gg r_e$$

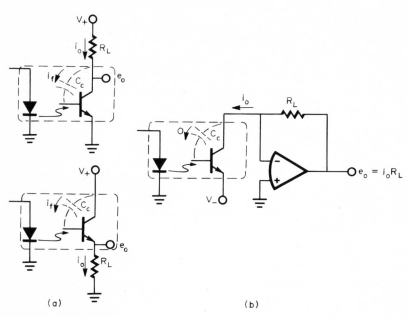

(a)   (b)

Fig. 2.25 Speed-limiting C_c feedback on phototransistors is eliminated by removing voltage swing with an operational amplifier.

where r_e is the dynamic emitter resistance.   Since the base is open-cir-cuited, this feedback current must all flow into the base region.   The re-sult is maximum ac feedback, which is not generally encountered with conventional transistor circuits having lower impedance base biases. Only when a conventional transistor has a current source base drive does it encounter this worst-case response condition.

Fortunately, this response limit can be removed by connecting an operational amplifier around the load resistor, as shown in Fig. 2.25b. Since feedback reduces the amplifier input signal voltage to zero, the sig-nal swing of the transistor is transferred to the amplifier output.   Then, C_c has no signal voltage and produces no shunting feedback current to the base.   Now frequency response will be determined primarily by the amplifier.

Due to the nonlinear transfer characteristic of the LED-phototransistor coupling, a high level of distortion is introduced by this signal transmis-sion.   Further, the high temperature sensitivity of the coupling efficiency creates a large gain drift with temperature.   To remove the distortion and gain drift, a compensating, nonlinear, temperature-sensitive feed-back can be supplied.   For best compensation this feedback is supplied through a similarly nonlinear and temperature-sensitive LED-photo-transistor coupling.   An isolated voltage amplifier using this technique is shown in Fig. 2.26.   In this circuit the output amplifier converts the transistor current to an output voltage as described above.

However, in this case the current supplied to the transmitting LED is controlled by a feedback LED-phototransistor coupling.   The two diodes are supplied the same level of current, and it will be that current level which makes the feedback phototransistor accept the signal current from e_i and the bias current from V_+.   This transistor current is essentially independent of the nonlinearity and thermal variations of the feedback LED-phototransistor coupling.   If the two diode-transistor sets are matched, then the same current will be developed in the output photo-transistor.   This produces an output voltage of

$$e_o = \frac{R_2}{R_1}\, e_i \qquad \text{if } V_+' = V_+$$

By accurately matching and heat-sinking the LED-phototransistor sets, an order-of-magnitude reduction in distortion and gain drift can be at-tained.   Some correction for mismatch can be achieved through adjust-ment of the diode current-setting resistors labeled R_4.

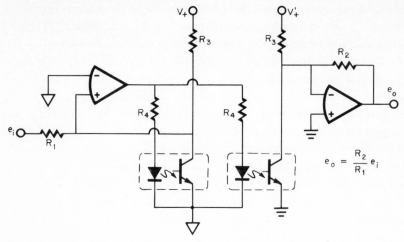

$$e_o = \frac{R_2}{R_1} e_i$$

Fig. 2.26 To remove distortion and gain drift from LED-phototransistor transmission, a matching optical coupling is connected as feedback.

In another application of linearizing feedback, isolating signal couplers can be connected with an operational amplifier to form a current amplifier with ground isolation. Such an amplifier is highly useful in the many process-control applications where the control signals are currents rather than voltages. In these applications many different monitors may be connected in series so that they all monitor a given current, but this series connection results in common-mode voltages at the various monitor inputs. Isolated couplers readily accommodate the common-mode voltages with no effect upon output signals.

For this current amplifier requirement where unipolar currents are adequate, the current-driven LED-phototransistor couplers are well suited. An isolation amplifier of this type is illustrated in Fig. 2.27. To compensate for the nonlinearity of the isolating coupler, feedback is applied through a matching coupler. An input current i_i will result in a current in the collector of Q_1 that will be matched by a feedback current from the emitter of Q_2. In order for this feedback current to be developed, the operational amplifier must supply an output current i_o through the load to the feedback coupler. This output current will approximately equal the input current i_i if the nonlinear responses of the two LED-phototransistor couplers match.

Several sources of error combine to produce a gain error of several percent in the above current amplifier. The main ones are coupler mismatch from several causes and the input bias current of the operational amplifier. Any mismatch in coupler current responses creates gain and linearity errors. Such mismatch can result from initial matching error, aging effects, or voltage bias differences. With age the coupling efficiencies of these isolating elements degrade significantly. Different collector-base bias voltages have a secondary effect on match which can be avoided by making the input reference divider 1:1 for equal bias voltages. Another, but generally negligible, error source is the input bias current of the operational amplifier. High noise from the LEDs can also degrade precision. Although the feedback loop forces an emitter current to equal a collector current, no error results, as a phototransistor has equal collector and emitter currents. The base current is supplied from the collector as leakage current.

2.5.2 Modulated carrier configurations   Precise signal transmission with ground isolation can also be attained by using modulated carrier techniques to avoid error from the nonlinearities of isolated couplers. The nonlinear transfer characteristics of isolated couplers distort amplitude but not frequency, and so either frequency modulation or pulse width modulation is used to transfer the signal across the coupler. With these modulation techniques and a square-wave carrier the coupler transmits only switching time information.

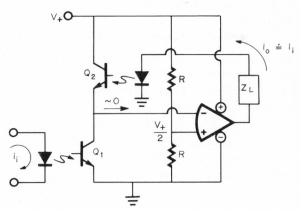

Fig. 2.27   A feedback linearized current amplifier with ground isolation is formed with matched LED-phototransistor couplers.

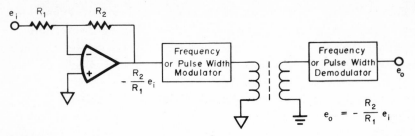

Fig. 2.28 An isolation amplifier can also avoid error from coupler nonlinearity by using a modulation technique that transfers only switching time information across the coupler.

Such an isolation amplifier can be built to be applied like an ordinary operational amplifier. With the isolation amplifier of Fig. 2.28 the feedback elements around the input buffer amplifier can be chosen to perform most of the functions of conventional operational amplifiers. This makes possible isolated voltage amplifiers, integrators, logarithmic amplifiers, multiplexers, controllers, and so forth. As shown, the output signal of the operational amplifier is modulated, then transferred by an isolated coupler and demodulated with respect to a new ground reference. While the coupler illustrated is a transformer, a variety of others could be used, including LED-phototransistor couplers, photoresistors, magnetoresistors, Hall-effect switches, and thermal couplers. The modulator and demodulator circuits are readily formed with operational amplifiers also, as described in Sec. 5.5.2, 5.5.3, and 6.2.2, and in other sources.[2]

For a modulated carrier isolation amplifier like the one described, the principal errors are the normal errors of the input operational amplifier combined with those of the modulator and demodulator. Essentially no error is introduced by the isolated coupler as long as its switching time is much less than the period of the carrier waveform. However, signal bandwidth is limited to frequencies well below that of the carrier at its modulated lower extreme. For that reason the carrier frequency is generally made as high as the switching times of the coupler, modulator, and demodulator permit. If the rise and fall times of a given switching waveform are identical, then the errors they create cancel. However, even if this error canceling condition exists perfectly, the combined switching times must be less than the minimum period of the modulated carrier.

REFERENCES

1. G. A. and T. M. Korn, *Electronic Analog and Hybrid Computers,* McGraw-Hill Book Company, New York, 1964.
2. G. E. Tobey, J. G. Graeme, and L. P. Huelsman, *Operational Amplifiers: Design and Applications,* McGraw-Hill Book Company, New York, 1971.
3. J. G. Graeme, Have a 50V Swing on a 30V Supply, *EDN/EEE.*
4. J. Brown, Current Input Differential Amplifiers with High Common-Mode Rejection, Proc. ISSCC, 1969.
5. J. G. Graeme, Op Amp Boosts Phototransistor Speed, *Electron. Design,* March 2, 1972.

3

SIGNAL CONDITIONERS

Operational amplifiers are applied in a wide range of signal conditioning circuits which alter the relationships of signals to time, to other signals, and to frequency. Some of these circuits include the integrators, differentiators, current sources, logarithmic amplifiers, multipliers, dividers, and active filters described in this chapter. Specialized integrators and differentiators which have expanded computation capability are described. Numerous voltage-controlled current sources which provide either unipolar or bipolar currents to either floating or grounded loads are presented. Modifications to the common logarithmic amplifier circuits are outlined. Operational amplifier circuits which perform multiplication and division through voltage-controlled resistances and by logarithmic techniques are explained. Techniques for adjusting phase, for voltage tuning of response characteristics, and for removal of dc offsets in active filters are described in the concluding section.

3.1 Integrators and Differentiators

Several variations on the basic integrator and differentiator circuits have been developed for performance alternatives. The basic circuits and their limitations are described in the first reference.[1] In this section the circuits described include specialized integrators, integrator RESET circuits, and specialized differentiators. For most of the specialized circuits additional computation is performed by the amplifier which performs the integration or differentiation.

3.1.1 Specialized integrators

By modifying the feedback network of an integrator circuit, several variations on the basic function are achieved. With symmetrical resistor/capacitor networks connected, as in Fig. 3.1, a differential integrator is formed. This circuit takes the difference between two signals and integrates the result as expressed in the accompanying equation. With this operation the output of a floating source can be integrated, and common-mode signals can be rejected. Common-mode rejection is determined by both the amplifier and the integrating components. A very accurate match between like components is required to preserve CMR, since this characteristic is directly related to the mismatch in the associated time constants. In terms of this mismatch the resistors and capacitors place a limit on CMRR of

$$\text{CMRR} \leq \frac{1 + \text{RCs}}{\Delta \text{RCs}}$$

where ΔRC is the difference between the time constants of the two networks. Performance is also limited by the input offset voltage and offset current of the amplifier in much the same way as with the basic integrator, and the associated error is

$$\epsilon = \frac{1}{\text{RC}} \int \text{V}_{\text{OS}} \, dt + \frac{1}{\text{C}} \int \text{I}_{\text{OS}} \, dt + \text{V}_{\text{OS}}$$

Another simple modification to an integrator for additional computa-

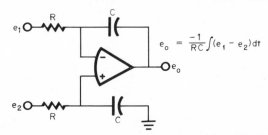

$$e_o = \frac{-1}{RC} \int (e_1 - e_2) \, dt$$

Fig. 3.1 A differential integrator is formed by adding a network identical to the feedback to the other input of the basic integrator.

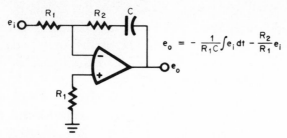

$$e_o = -\frac{1}{R_1 C}\int e_i\,dt - \frac{R_2}{R_1}e_i$$

Fig. 3.2 An added feedback resistor converts the simple integrator to a circuit which produces an amplified sum of input signal and its integral.

tion by the circuit is shown in Fig. 3.2. In this case a resistor is added in series with the feedback capacitor to form an augmented integrator. Computation capability is thus augmented to provide an output equal to the amplified sum of the input signal and its integral.   This is expressed by the equation of the figure and can be visualized by using superposition. With superposition, R_2 and C can be considered separately as feedback elements to define the two terms of the output equation.   Considering the resulting inverting amplifier and simple integrator circuits, the errors caused by input offset voltage and current can also be described. The result is an error of

$$\epsilon = \frac{1}{RC}\int V_{os}\,dt + \frac{1}{C}\int I_{os}\,dt + \frac{R_2}{R_1}V_{os} + I_{os}R_1$$

A similar integration is performed on a signal applied to the noninverting amplifier input of a conventional integrator.

A single operational amplifier can also produce a double integral when connected as in Fig. 3.3. Essentially the circuit is a two-pole, low-pass

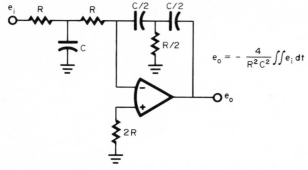

$$e_o = -\frac{4}{R^2 C^2}\iint e_i\,dt$$

Fig. 3.3   A double integration is performed by a two-pole, low-pass filter with coincident poles.

filter with one pole produced by each of the two T networks. If the components have accurately coinciding response poles, the circuit response becomes that of a double integration with respect to time. As expressed, the integrator gain is now related to the square of the resistor and capacitor values. This new gain and the double integration also act upon the input offset voltage and current of the operational amplifier to create an error of

$$\epsilon = \frac{4}{R^2 C^2} \iint V_{OS} \, dt + V_{OS} + \frac{4}{RC^2} \iint I_{OS} \, dt + \frac{4}{C} \int I_{OS} \, dt$$

3.1.2   Integrator reset
The output signal developed with an integrator is cumulative and is not removed when the input signal returns to zero. Since the feedback capacitor holds the output voltage in this manner, repeated integrations can build up the output to its saturation level. To avoid this the integrator must be periodically reset, unless it is used in a feedback loop which controls the output voltage. The basic modes of operation used to provide integrator reset are well described elsewhere.[2] Described here are a low-leakage RESET switch and an automatic RESET circuit.

As represented in Fig. 3.4, a low-leakage RESET switch can be formed with two MOSFETs. Reset to zero output voltage is achieved by applying a negative pulse to the gates of the FETs to short-circuit the feedback

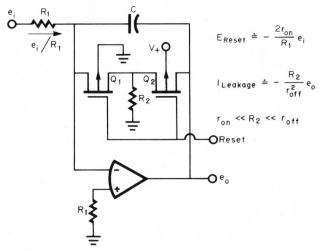

Fig. 3.4   Two FETs form an integrator RESET switch with greatly reduced switch leakage current in the input.

capacitor. In this condition the capacitor discharges through the FETs, and the input signal current e_i/R_1 also flows through these switches. The capacitor discharges to the low voltage maintained across the switch ON resistances r_{ON} by the current e_i/R_1. As a result, the capacitor voltage is not reset to exactly zero, and a RESET error voltage is induced which for $r_{ON} \ll R_2$ is

$$E_{RESET} \doteq - \frac{2r_{ON}}{R_1} e_i$$

Two FET switches are used to achieve a dramatic reduction in leakage current when the switches are at OFF and the integrator is operating. This leakage current reduction technique parallels the leakage decoupling circuit used with feedback limiters.[1] It is the leakage current from Q_1 which creates an error, since this current adds to the input signal current e_i/R_1. If Q_1 alone were used and were connected directly across C, the leakage inducing voltage across the FET would be the entire output voltage. In many cases the resulting leakage current would even be greater than the input bias current of the amplifier and would be the major source of error. However, the addition of Q_2 and R_2 decouples this large leakage current from the input. Now the voltage across Q_1 will be very low, so that the leakage current reaching the input is small. One side of Q_1 is held at zero voltage at the amplifier input, and the other side is held at a voltage created by the leakage current of Q_2 flowing in R_2. This second voltage will very nearly equal $(e_o/r_{OFF})R_2$, where r_{OFF} is the OFF resistance and it is assumed that the voltage across Q_2 is essentially e_o. Then the voltage across Q_1 is $(e_o/r_{OFF})R_2$, from the above. This develops a leakage current to the input related to the OFF resistance of Q_1, which can be expressed by

$$I_{leakage} = - \frac{R_2}{r_{OFF}^2} e_o$$

The result is an input leakage current reduction by a factor of r_{OFF}/R_2.

It is often desirable to reset an integrator when its output reaches a certain voltage level, such as in some integrating digital voltmeters. A nonzero RESET voltage is also frequently required. These reset characteristics are provided by the automatic RESET circuit of Fig. 3.5. In this case the circuit of Fig. 3.4 is modified to include a comparator A_2 for the RESET drive. Since the comparator is driven by the integrator out-

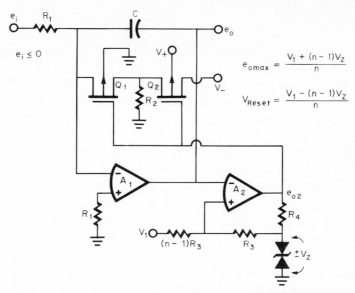

$$e_{o\,max} = \frac{V_1 + (n-1)V_Z}{n}$$

$$V_{Reset} = \frac{V_1 - (n-1)V_Z}{n}$$

Fig. 3.5 Automatic RESET to any voltage is provided by a comparator driven from the integrator output.

put, reset will occur when this output reaches the trip voltage of the comparator. This defines the maximum output voltage $e_{o\,max}$ shown. When $e_{o\,max}$ is reached, the comparator output switches negative to turn the FETs to ON and discharge C. Discharge continues to some lower voltage as determined by the hysteresis of the comparator. Hysteresis is developed by the comparator positive feedback through R_3 and permits reset to any voltage below $e_{o\,max}$. Both the RESET voltage V_{RESET} and $e_{o\,max}$ as expressed in Fig. 3.5 are set by choosing reference voltage V_1, zener voltage V_Z, and the constant n of the feedback.

RESET error is reduced by the feedback of this technique. With the preceding circuit a RESET error voltage remained on C due to the signal current flow in the switch ON resistances. However, in this case the comparator compares the output against a reference during the RESET period. As a result, C will be discharged until e_o equals the desired level independent of switch voltage drops. Some RESET error remains due to the turnoff time of the switches, which causes the capacitor to discharge too far. So that the added discharge is small during the turnoff time, the discharge rate can be limited by resistance in series with the switch Q_2. Another RESET error will be determined by the errors of the comparator and will be

$$E_{\text{RESET}} = V_{\text{OS2}} + I_{\text{B2}}\left(1 - \frac{1}{n}\right)R_3 + \frac{e_{o2}}{A_2}$$

where V_{OS2}, I_{B2}, and A_2 are the input offset voltage, input bias current, and open-loop gain of the comparator. This error can be removed by making compensating adjustments in the comparator trip points.

3.1.3 Specialized differentiators

Variations on the basic differentiator circuit, paralleling those of Sec. 3.1.1 for an integrator, provide a differential differentiator and an augmenting differentiator. Also described in this section is an indirect approach to differentiation which improves upon the noise problem characteristic of differentiators. The differential differentiator of Fig. 3.6 is formed by adding to the noninverting input a network which is identical to the feedback network. In each network a gain-limiting resistor R_1 is added to the basic elements in order to ensure frequency stability.[1] Below the frequency of this gain limit the response of the circuit is approximated by the expression in the figure. As expressed, the differential differentiator takes the difference between two signals and differentiates the result. In this way the output of a floating source can be differentiated without its common-mode signal.

Common-mode signals introduce one of the errors of this circuit. The common-mode rejection is limited by mismatch in the attenuations of the two networks and by the CMR of the operational amplifier itself. For the CMRR limit imposed by the networks, an expression can be written from the fractional mismatch and the circuit gain which reduces to

$$\text{CMRR} \leq \frac{\Delta RCs}{1 + RCs}$$

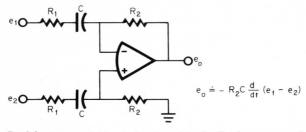

$$e_o \doteq - R_2C \frac{d}{dt}(e_1 - e_2)$$

Fig. 3.6 A network identical to that of the feedback is connected to the noninverting input of a differentiator to form a differential input.

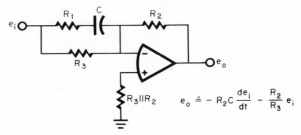

Fig. 3.7 An added input resistor R_3 converts the simple differentiator to a circuit which produces a signal equal to the amplified sum of the input signal and its derivative.

Gain error also limits the accuracy of differentiator circuits such as this. Ideally, a differentiator would display a continuously increasing gain with increasing frequency, which cannot, of course, be attained. Gain-bandwidth limitations of the operational amplifier greatly limit differentiator response. When the stabilizing gain-limit resistor R_1 is used, the differentiator response approximation ends at $f = 1/2 \, \pi \, R_1 C$. Operational amplifier dc errors also create an error, and this will be an output voltage of

$$\epsilon = V_{OS} + I_{OS} R_2$$

The computation performed by a differentiator can also be augmented to include an output signal term from an inverting amplifier. This augmented differentiator,[3] as shown in Fig. 3.7, varies from the basic differentiator by an added resistor R_3. Using superposition, it is seen that this input element alone would give inverting amplifier operation, while R_1 and C alone provide differentiator operation. Similarly, the error of this circuit is the sum of the errors of the associated inverting amplifier and differentiator circuits. A similar differention is performed on a signal applied to the noninverting amplifier input of a conventional differentiator.

A major problem encountered with differentiator circuits is high noise. This results from the increasing gain of the differentiator frequency response and the resulting high-gain amplification of the amplifier noise. With this differentiator gain the equivalent input noise voltage of the amplifier e_n creates an output of

$$e_o = RCse_n$$

To reduce this noise an indirect method of computing the derivative of a signal can be used, as in Fig. 3.8. In this case the derivative of the signal

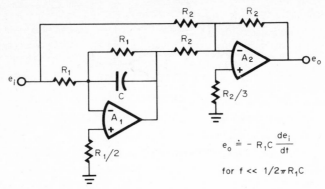

$$e_o \doteq - R_1C \frac{de_i}{dt}$$

$$\text{for } f \ll 1/2\pi R_1 C$$

Fig. 3.8 The derivative of a signal can also be computed by an indirect approach to avoid the high noise of the basic differentiator.

is derived from the signal and its integral. By analysis the response of the circuit is found to be

$$e_o = \frac{-R_1 Cs}{1 + R_1 Cs} e_i$$

This response is of the same form as that of the simple differentiator with a gain-limiting resistor R_1 equal to the feedback resistor. Once again the denominator term limits the differentiator response approximation, but for $f \ll 1/2\pi R_1 C$,

$$e_o(s) \doteq - R_1 Cs \, e_i \qquad \text{or} \qquad e_o(t) \doteq -R_1 C \frac{de_i}{dt}$$

However, the amplifier noises are not multiplied by the high differentiator gain. Instead the noise of A_1 is integrated, resulting in a noise that decreases rather than increases with increasing frequency. The noise of the integrator is combined with that of the inverter to develop an output of

$$e_o = \frac{e_{n1}}{1 + R_1 Cs} + e_{n2}$$

3.2 Controlled Current Sources

Most electronic instrumentation is composed of circuits which produce a signal voltage from a control voltage. However, signal currents derived from control voltages can sometimes provide more straightforward solutions for instrumentation requirements. Such voltage-controlled current sources are useful in testing and for driving certain loads. In transistor testing, as described in Chapter 6, controlled current sources provide simple, programmable current biasing. Resistance measurement is simplified with a current rather than a voltage test signal, as contact resistance will not affect the current supplied from a current source. Current output is also needed for meter drive and dc torque motor drive. For such motors the torque developed is proportional to the drive current. Operational amplifiers can be used to form precise controlled current sources because of the high-gain feedback possible with these amplifiers. Described in this section are a variety of such current sources, with both unipolar and bipolar outputs, intended for both floating and grounded loads. Additional voltage-to-current converters are described in the first reference.[1]

3.2.1 Unipolar output current sources

For applications requiring only one polarity of output current, high-performance current sources can be formed with only an operational amplifier, a transistor, and a resistor. Very high output resistances and output currents can be attained with these circuits, and the loads do not have to be floated. The most straightforward form is shown in Fig. 3.9a. This configuration resembles the simple transistor current source in that the transistor base is biased to fix a voltage on the emitter resistor. In this case the resistor voltage is sensed by one input of the amplifier, and feedback forces the voltage at that point to equal the input voltage. As a result, the signal voltage on R is precisely controlled by e_i without error from the emitter-base voltage of the transistor. The resulting controlled current in R essentially all flows in the transistor emitter, with only a very small current drawn by the sensing input of the amplifier. However, the collector current which flows from the output differs from this controlled emitter current due to a transistor alpha that is less than unity. This results in an output of

$$i_o = a\, \frac{V_+ - e_i}{R}$$

A Darlington pair can be used in place of the single transistor to greatly reduce this error.

Except for the alpha error the current source peformance is quite good. Even the basic error due to alpha can be compensated by an

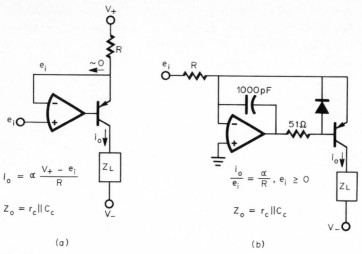

Fig. 3.9   A basic transistor current source is controlled for accuracy and high output resistance by the high-gain feedback of an operational amplifier.

adjustment in resistor value.   This leaves only a small linearity error due to the variation in alpha with current level.   Output current can be quite high since it is not supplied through the amplifier or from the input.   High output impedance is developed by the reverse-biased collector-base junction.   Since the voltage swing on the load is absorbed by the transistor collector-base junction, the output impedance equals the impedance of that reverse-biased junction, or

$$Z_o = r_c \| C_c$$

The output resistance r_c equals β/h_{oe} and will be of the order of 10 to 100 MΩ while the output capacitance C_c will be typically 3 pF.   In addition, with the load voltage swing on the transistor rather than the amplifier, the load voltage slewing rate is not limited by the amplifier.

A significant limitation of the circuit of Fig. 3.9a is that the output current, as expressed with the circuit, is sensitive to positive supply variations.   To avoid the sensitivity the input signal e_i can be referenced to V_+ instead of to ground.   Alternatively, a grounded source can drive the emitter resistor directly to avoid supply sensitivity, as in Fig. 3.9b.   Although this latter connection lowers input impedance and supplies the output current from the signal source, performance is otherwise the same as before.   In this case the operational amplifier feedback holds the transistor emitter at very nearly zero volts so that the input signal is impressed on R, giving $i_o = \alpha e_i / R$. The added phase compensation shown is required for stability when e_i approaches zero, causing the transistor emitter impedance to become high near turnoff. A diode clamp is also added, and this prevents reverse breakdown of the emitter-base junction.

In higher-precision applications the alpha error of the previous circuits is significant. This error can be avoided by using FETs in analogous circuits, as shown in Fig. 3.10. Output resistance is also improved. Circuit operation parallels that described above for the bipolar transistor circuits with feedback controlling the source resistor voltage. Essentially all the resulting controlled current flows in the FET source. In this case the controlled current is passed to the load with no transmission loss like that of the base current before, and for the two circuits

$$i_o = \frac{V_+ - e_i}{R} \quad \text{and} \quad i_o = \frac{e_i}{R}$$

However, the maximum current which can be supplied to the load through the FET is I_{DSS}. High output currents are not as easily attained unless current is boosted, as described in the next section.

Output impedance is boosted with the FETs since load voltage

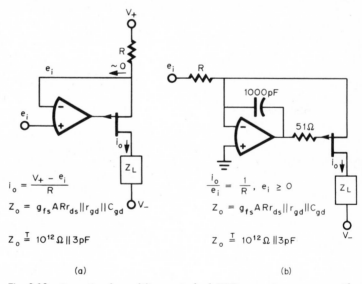

(a) (b)

Fig. 3.10 Operational amplifier control of FET current sources provides high accuracy and extremely high output resistance.

swing is impressed on the gate-drain junction. The FET junction presents the same high reverse impedance from gate to drain as is commonly seen from gate to source. Feedback corrects for the output impedance shunt normally presented by the drain-source resistance r_{ds}, and the resulting output impedance is

$$Z_o = g_{fs} A r_{ds} \parallel r_{gd} \parallel C_{gd} \stackrel{T}{=} 10^{12}\ \Omega \parallel 3\ pF$$

Unfortunately this ultrahigh $10^{12}\ \Omega$ is usable only at very low frequencies, since it breaks with the 3 pF output capacitance at only 0.05 Hz. Output guarding reduces this capacitive shunting, as described in Sec. 3.2.3. The output impedance is also decreased with frequency by the fall in amplifier gain and the associated loop gain decrease.

Current sources similiar to those described above can be formed with the emitter or source as the output terminal instead of the collector or drain. This approach is illustrated in Fig. 3.11 for bipolar transistors. While it might be expected that the emitter would present a low output impedance, the operational amplifier feedback produces a high impedance output. Voltage changes on the load are tracked by the amplifier output in order to maintain a constant output current. This transfers load voltage swings to the transistor collector-base junction and results in an error signal at the amplifier inputs equal to this swing divided by the open-loop gain. From this error signal the output impedance is

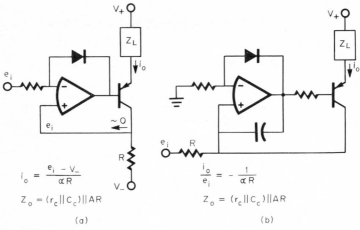

$$i_o = \frac{e_i - V_-}{\alpha R}$$

$$Z_o = (r_c \| C_c) \| AR$$

(a)

$$\frac{i_o}{e_i} = -\frac{1}{\alpha R}$$

$$Z_o = (r_c \| C_c) \| AR$$

(b)

Fig. 3.11 Feedback control of an emitter follower results in a current source with added feedback gain from the transistor.

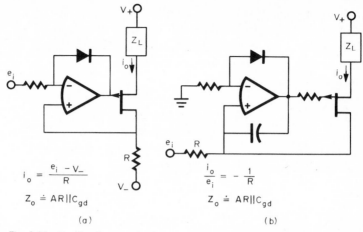

$$i_o = \frac{e_i - V_-}{R}$$

$$Z_o \doteq AR \| C_{gd}$$

(a)

$$\frac{i_o}{e_i} = -\frac{1}{R}$$

$$Z_o \doteq AR \| C_{gd}$$

(b)

Fig. 3.12 Feedback control of a source follower also produces a current source with added feedback-loop gain.

defined by

$$Z_o = (r_c \| C_c) \| AR$$

In these circuits the transistor adds gain to the feedback loop and reduces that required from the amplifier.

Feedback now controls the transistor collector voltage, maintaining near-zero voltage between the amplifier inputs. This establishes a collector current related to the input signal and results in an emitter current to the load. Once again the transistors introduce an alpha error as indicated by the output current expressions

$$i_o = \frac{e_i - V_-}{aR} \quad \text{and} \quad i_o = -\frac{e_i}{aR}$$

Since the load voltage swing is tracked by the amplifier, the slewing rate of this voltage is limited by that of the amplifier. A feedback diode clamp and resistor are added to each circuit to prevent latching from phase reversal accompanying transistor saturation. For inductive loads additional clamping should be used to avoid reverse breakdown of the emitter-base junction.

Once again the alpha error of bipolar transistors can be avoided with FETs, as shown in Fig. 3.12. This results in current sources analogous to those of the last figure, with greater accuracy but with output current limited to I_{DSS} or less.   Output currents are

$$i_o = \frac{e_i - V_-}{R} \quad \text{and} \quad i_o = -\frac{e_i}{R}$$

High output impedance is developed by feedback and the high reverse impedance of the FET junction.   With a feedback error signal at the input equal to the output voltage swing divided by the voltage gain, the circuits have the output impedance shown.   This impedance decreases with frequency as a result of the decreasing gain and the shunting of the gate-drain capacitance.   Since the amplifier follows the load voltage swing, the slewing rate is limited to that of the amplifier.

3.2.2 Improving unipolar output characteristics   Simple circuit additions can improve the output impedances of the bipolar transistor-controlled current sources and can increase the output currents of the FET-controlled current sources.   Improved output impedance is achieved by adding cascode bias, as shown in Fig. 3.13 for the circuit of Fig. 3.9a. With the addition of the cascode-biasing FET, the collector-base voltage of the current source transistor equals the gate-source voltage of the FET.   Load voltage variations are not impressed on the collector-base

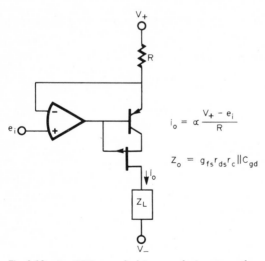

$$i_o = \alpha \frac{V_+ - e_i}{R}$$

$$Z_o = g_{fs} r_{ds} r_c \| C_{gd}$$

Fig. 3.13   An FET cascode bias greatly increases the output impedance of a bipolar transistor-controlled current source by essentially removing load voltage swing from the current source transistor.

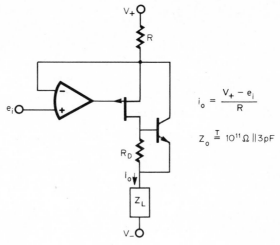

$$i_o = \frac{V_+ - e_i}{R}$$

$$Z_o \stackrel{\cdot}{=} 10^{11}\,\Omega \,\|\, 3\,\mathrm{pF}$$

Fig. 3.14 A bipolar transistor connected as shown multiplies the transconductance of the FET to permit output currents greater than I_{DSS}.

junction, which previously set the limit on output resistance. Instead, the load voltage swing is essentially absorbed by the FET. This swing does not directly create an output current change with the r_{ds} of the FET, since the FET must pass whatever current is supplied by the bipolar transistor. To assure this condition the bipolar transistor drives the FET source to whatever voltage is required for transmission of the collector current. It is this source drive voltage that is the remnant of output voltage swing on the collector-base junction, and this swing is reduced by a factor of $g_{fs}r_{ds}$. Then, the output impedance is boosted by the same factor, resulting in

$$Z_o = (g_{fs}r_{ds}r_c) \,\|\, C_{gd}$$

A disadvantage of the FET cascode bias is that output is limited to I_{DSS} or less, as with the FET-controlled current sources before.

To remove the output current limitation of FET-controlled current sources or of the above cascoded circuit, a transconductance-multiplying transistor can be added. Such a transistor is shown in Fig. 3.14 for the circuit of Fig. 3.10a. With this connection the output current is not limited to I_{DSS}, since additional current is supplied by the bipolar transistor. A small drain current from the FET results in a larger current from the bipolar transistor, with the total current equaling that supplied through R. The error of this current is only the gate-drain leakage current of the FET plus the input bias current of the operational amplifier. Transconductance is multiplied as set by R_D and the dynamic emitter resistance r_e. By these resistances the multiplying factor is set at approximately R_D/r_e, with a maximum limit equal to the transistor beta. Output impedance is reduced by this same factor but remains quite high as expressed. Overall performance of this current source is outstanding, as it provides high output current from a high impedance with only a small error.

3.2.3 Special-purpose unipolar current sources

For more specialized current source requirements several variations on the preceding unipolar circuits can be made. Such variations permit grounded loads, remove the alpha error with bipolar transistors, provide multiple outputs, guard the output, and provide an output voltage monitor without shunting the output. A load returned to ground, rather than to a power supply as before, may be desirable to avoid dc level shifting. With the preceding unipolar circuits the load return to a power supply provided voltage bias for the transistor. Alternately, the transistor can be biased away from ground as in Fig. 3.15. In this case the load voltage can swing below ground without saturating the output transistor since the base is biased below ground.

Operation with this bias connection also results in other performance changes. First, the alpha error encountered with other bipolar transistor-controlled current sources is avoided because of the added transistor. If the transistors are matched, the common-emitter-base bias

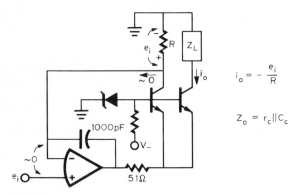

Fig. 3.15 By biasing the current source transistor below ground, a grounded load can be driven without saturating the transistor.

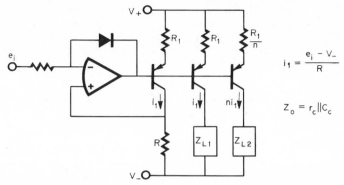

$$i_1 = \frac{e_i - V_-}{R}$$

$$Z_o = r_c \| C_c$$

Fig. 3.16 Multiple current source outputs are achieved with common biasing of current source transistors.

will result in equal collector currents. Then the output current will equal the current generated in R by e_i, or $i_o = -e_i/R$. The alpha errors of the two transistors cancel. The accuracy of this error compensation depends upon a match in alphas and emitter-base voltages over a wide range of currents, which is best achieved with monolithic dual transistors. With discrete transistors, matching emitter resistors should be added to ensure equal currents, as a mismatch becomes self-increasing by creating a thermal mismatch. Also, this technique does limit the output current, which must be supplied from the operational amplifier output along with the current in the matching transistor.

Multiple, weighted current source outputs are useful in such applications as digital-to-analog converters. With the technique of Fig. 3.16 any number of outputs with arbitrary weighting can be produced. The bias voltages impressed on each of the emitter resistors will be nearly equal. These voltages will differ by only the differences in emitter-base voltages, which are reduced by transistor matching and by paralleling transistors in the higher-current sections. If these voltage differences are negligible, the emitter currents can be ratioed as shown, using different values of emitter resistance. Output currents have the same ratioing if the transistor alphas are matched, and the alpha transmission error is removed as in the previous circuit. The basic feedback control circuit used here is the same as described earlier for Fig. 3.11. From this feedback the magnitude of the basic current i_1 is precisely controlled as expressed with this new circuit. Output impedance, as expressed, is the same as before, and the clamp diode is again required to avoid circuit latching when the feedback transistor saturates.

Very high output resistances can be attained with operational amplifier current sources, but even small output capacitances can greatly limit the utility of this high resistance. An output resistance of 10^{12} Ω shunted by only 3 pF has a response pole at 0.05 Hz. Capacitance from a load voltage monitor would introduce an even lower-frequency pole. To extend the useful range of high output impedances, the capacitance shunting the output can be reduced with a guard shield.[4] Both the guard shield and a load voltage monitor must be driven by a signal which tracks the load voltage. Such a signal is generated with A_1 in Fig. 3.17 by means of a feedback loop. Feedback forces the voltage between the inputs of A_2 to zero and establishes the output of A_1 at the load voltage for the desired drive signal. The amplifier output provides a low impedance drive to the shield, any voltage monitor, and the stray capacitance at the output. With the driven shield the output is isolated from stray capacitances and the monitor capacitance. No output shunting results from the capacitance between the shield and output, since they have the same voltage swing.

The output current is also set by the feedback loop, beginning with the comparison made by A_2. With the inputs of this operational ampli-

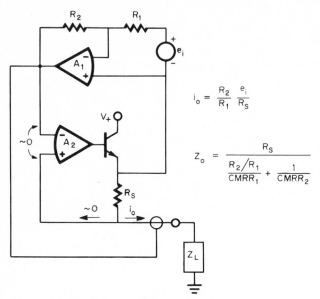

$$i_o = \frac{R_2}{R_1} \frac{e_i}{R_S}$$

$$Z_o = \frac{R_S}{\dfrac{R_2/R_1}{\mathrm{CMRR}_1} + \dfrac{1}{\mathrm{CMRR}_2}}$$

Fig. 3.17   With the two amplifier current sources shown, a voltage equal to that at the output is derived for driving a guard shield and any voltage monitor.

fier connected to the sense resistor R_S and to the output of A_1, A_2 compares the voltages of the two with respect to common. At the output of A_1 the voltage drop from the emitter is $-e_i R_2/R_1$, and the feedback will establish an equal voltage on R_S to set i_o at

$$i_o = \frac{R_2}{R_1} \frac{e_i}{R_S}$$

From the high-gain feedback of the operational amplifiers a highly precise current is established within the accuracies of the resistors. The sensitivity of this current to load voltage changes is held low by the common-mode rejections of the amplifiers. Such voltage changes appear as common-mode signals to the amplifiers, and the entire current source floats above the load voltage. Common-mode errors result in an output current sensitivity defined by an output impedance of

$$Z_o = \frac{R_S}{[(R_2/R_1)/CMRR_1] + (1/CMRR_2)}$$

3.2.4 Bipolar output current sources for floating loads
All the previously described current sources provided unipolar output currents as required in a variety of applications such as constant current sources. However, more general-purpose controlled current sources can be formed with operational amplifiers to provide both polarities of output currents. Several such bipolar voltage-to-current converters are described elsewhere.[1] Often the load to be supplied by a current source can be floated rather than grounded, and this condition permits the use of several simplified bipolar-controlled current sources described in this section.

In the most basic current source configuration the load is simply connected in the feedback loop of the operational amplifier.[1] Then the load current is the well-controlled feedback current supplied in response to an input signal. This current is returned to the signal source in inverting amplifier connections commonly required for chopper-stabilized and some wideband operational amplifiers. As a result, a current gain of only unity is achieved with these inverting-only amplifiers in the simplest current source configuration. Greater current gain can be developed by adding a current divider to the feedback network as in Fig. 3.18. Now only a fraction of the load current is fed back to the signal source through the divider formed by R_2 and R_3. A current gain of $1 + R_2/R_3$ is

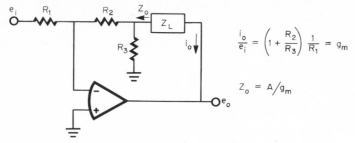

$$\frac{i_o}{e_i} = \left(1 + \frac{R_2}{R_3}\right)\frac{1}{R_1} = g_m$$

$$Z_o = A/g_m$$

Fig. 3.18 By adding a resistor divider to the feedback loop, R_2 and R_3, the load current is boosted above that supplied from the signal source in this current source configuration.

achieved to reduce the loading on the signal source and give a transconductance of

$$\frac{i_o}{e_i} = g_m = \left(1 + \frac{R_2}{R_3}\right)\frac{1}{R_1}$$

However, the feedback divider also increases the output voltage swing required to supply the load current and reduces the feedback-loop gain. From the decreased loop gain a drop occurs in the output impedance presented by the current source to the load, resulting in an output impedance of $Z_o = A/g_m$.

A second current source configuration suitable for inverting-only operational amplifiers provides an even greater reduction in source loading where the source can be floated. In this case both the source and the load are connected in the feedback loop in Fig. 3.19. Alternatively, a grounded signal may drive the noninverting input. In either case the current drain on the source is only the input bias current of the amplifier, making this approach well suited for standard cell sources. Further, the input impedance presented to the source by the amplifier is exceptionally high, since the signal source is connected directly to an amplifier input and since feedback essentially removes the signal swing from this input. The result is an input impedance of $Z_I' = AZ_I$ for the case shown, where A and Z_I are the open-loop gain and input impedance, respectively. Stray leakages limit Z_I' to around 10^{12} $\Omega \parallel 1$ pF.

Feedback controls the load current by reducing the amplifier differential input voltage to nearly zero so that the voltage on the sense resistor R_S equals the input signal e_i. The associated current e_i/R_S essentially all flows in the load if the amplifier input current is negligible. As a result, the output current is accurately and conveniently controlled by a single resistor R_S to be $i_o = e_i/R_S$. High-gain feedback removes most of the signal error associated with the amplifier, as expressed by the output impedance presented to the load. Load voltage changes produce small input voltage changes, and thereby current changes, to define the current source output impedance as $Z_o = AR_S$. However, this high output impedance is presented only to loads connected to the sense resistor R_S, and not to output loads returned to ground. Any grounded load, such as a voltage monitor, does not return a feedback current to the sense resistor to change the output voltage. Low output impedance to a grounded monitor permits use of the output for an indirect measure of the load voltage, which equals $e_o - e_i$.

A direct measure of the load voltage, without output impedance reduction by a monitor, is provided by the current source of Fig. 3.20. Any monitoring load connected as shown by the meter will be driven by the outputs of the two operational amplifiers with a voltage equal to the load voltage. None of the current drawn by this monitor will flow in the sense resistor R_S because of the buffering provided by voltage follower A_2. Thus, the addition of a monitor will not affect the load current, so that the output impedance presented to the load is unchanged. Only the current through the load, less the small input current of A_2, flows in the sense resistor to generate a voltage for feedback. This voltage is compared with the input signal e_i on the feedback resistors R_1 and R_2. When $i_oR_S/R_2 = -e_i/R_1$, the input current and voltage of A_1 are essentially zero and the feedback loop is in equilibrium. Then

$$i_o = -\frac{R_2 e_i}{R_1 R_S}$$

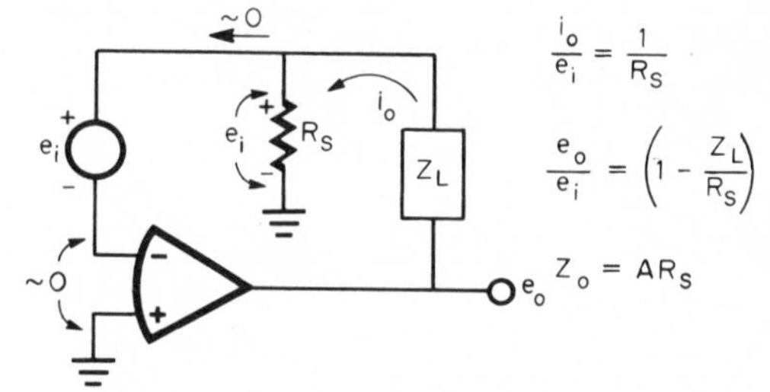

$$\frac{i_o}{e_i} = \frac{1}{R_S}$$

$$\frac{e_o}{e_i} = \left(1 - \frac{Z_L}{R_S}\right)$$

$$Z_o = AR_S$$

Fig. 3.19 When both the load and the source may be floated, the controlled current source shown will provide very light source loading even with inverting-only operational amplifiers.

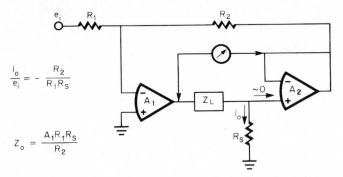

$$\frac{i_o}{e_i} = - \frac{R_2}{R_1 R_S}$$

$$Z_o = \frac{A_1 R_1 R_S}{R_2}$$

Fig. 3.20 A voltage-follower buffer A_2 in the feedback loop of this current source permits connection of a load voltage monitor without shunting the current source output impedance.

Also established by the feedback is a high output resistance presented to the load by the current source. Finite gain in A_1 limits the output impedance to that shown. The frequency stability of this feedback loop is affected by phase shift in the load and may require a compensating bypass capacitor between the two amplifier outputs.

3.2.5 Bipolar output current sources for grounded loads   Grounded loads are somewhat more difficult to supply from bipolar-controlled currents. Circuits providing this type of output are described elsewhere[1] and in this section for either floating or grounded sources. When the signal source can be floated, the controlled current source of Fig. 3.21 can be applied. This configuration is analogous to that of Fig. 3.19 except that the load is grounded and the operational amplifier must have a noninverting signal input. Once again the signal source is very lightly loaded by the operational amplifier, as is desirable for sources such as standard cells. The source supplies the small input bias current of the amplifier and is loaded by a high input impedance of $Z_{Icm}R_S/(R_S + Z_L)$.

Feedback controls the output current by forcing the differential input voltage of the amplifier to essentially zero so that the voltage on the sense resistor R_S equals the signal voltage e_i. The resulting current e_i/R_S nearly all flows into the load, as the amplifier input current drawn through the source is small. Note that this condition is disturbed by any source impedance to ground, such as biasing, since any such impedance draws current from the sense resistor. From the output current flow in the load a voltage is developed on which the current source floats. This voltage is a common-mode signal to the operational amplifier and results in common-mode errors that create current error. Output impedance expresses this sensitivity to load voltage in terms of the amplifier common-mode rejection ratio as

$$Z_o = (A \parallel CMRR)R_S$$

When the signal source cannot be floated, or when the source needs a biasing current, bipolar current drive of grounded loads can be provided by the circuit of Fig. 3.22. This circuit is essentially the unipolar circuit of Fig. 3.10 expanded for bipolar operation by using two output FETs. Feedback forces the FET source resistors to have voltage variations equal to the input signal e_i. On one of the resistors the direction of the signal increases current, and on the other resistor it creates the identically opposite change. The current changes unbalance the drain currents to develop an output current of $i_o = -2e_i/R_S$. To permit this signal current, the amplifier output must have a voltage swing capability of $5.7e_i$ with the divider resistor values shown. Output current is limited to I_{DSS} unless boosted as before in Fig. 3.14.

Errors in this circuit are primarily the signal error associated with finite output impedance and a dc offset error. Once again feedback greatly increases output impedance over that of a simple FET current source. Load voltage variations on the FETs do not result in source current changes with the r_{ds} of the FETs since feedback controls the source currents. Instead the load voltage swing can change output current by only the amount shunted from drain to gate through the reverse junction impedance. Output impedance is then approximated by $Z_o = 10^{12}$

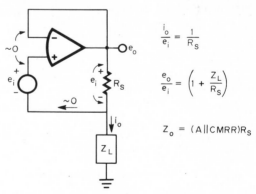

$$\frac{i_o}{e_i} = \frac{1}{R_S}$$

$$\frac{e_o}{e_i} = \left(1 + \frac{Z_L}{R_S}\right)$$

$$Z_o = (A \parallel CMRR)R_S$$

Fig. 3.21 For a floating source with no bias return to ground and a grounded load, a simple current source is available.

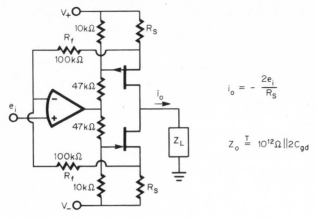

Fig. 3.22 Operational amplifier control of opposing FET current sources results in a bipolar current source for use with grounded load and source.

$\Omega \parallel 2C_{gd}$. The other basic error of this circuit is an output offset current resulting from mismatches in this balanced structure. Unequal power-supply voltage magnitudes, source resistors R_S, or feedback resistors R_f will result in significant offset error. Much smaller errors result from the input offset voltage and input bias current of the operational amplifier. Only a very small offset is produced by mismatched voltage dividers at the amplifier output, since the effect of this mismatch is divided by the amplifier gain.

3.3 Logarithmic Amplifiers

The exponential current-voltage relationship of a semiconductor junction is one of the most predictable characteristics found in electronics. With this precise exponential relationship, logarithmic amplifiers handling many decades of signal level can be formed, using operational amplifiers as previously described.[1] Logarithmic amplifier accuracy is degraded at low current levels by leakage currents and at high current levels by the bulk resistance of the semiconductor junction. At higher frequencies accuracy is further limited by the phase compensation required in the common logarithmic amplifier configuration. In this section techniques are described for improving accuracy through compensation of bulk resistance errors and through use of a nonlinear phase compensation.

3.3.1   Compensation of bulk resistance error

To make use of the exponential current-voltage characteristic of a junction, it is desirable to monitor only that portion of the junction voltage which bears this relationship. However, the junction current flow also creates a voltage drop with the bulk resistance of the junction, and this added voltage bears a linear relationship to the current. At higher current levels this linear term produces a significant error, but this term can be compensated. The network for compensating this error is shown in Fig. 3.23 on an elementary logarithmic amplifier circuit. In this circuit the $R_1 D_1 R_2$ network provides compensation for the logarithmic $R_1 D_2$ network. Both diodes are represented by an ideal logarithmic diode and a separate bulk resistance r_B.

As indicated, the basic output voltage e_o differs from the desired junction voltage v_f by the bulk resistance error term $i_f r_B$. By subtracting an equal signal voltage from the output, the error is removed. A signal current nearly equal to i_f flows in the compensation network if $R_2 \ll R_1$. Then by making $R_2 = r_B$, the required compensating signal is generated to give $e_o' \doteq - v_f$, as desired. In this way the error from r_B can be reduced an order of magnitude. The new output terminal cannot be loaded significantly, however, without disturbing the error compensation.

An analogous technique can be applied to remove the error from r_B in an antilog amplifier,[5] as shown in Fig. 3.24. Here the elementary antilog amplifier is modified by the addition of the $R_1 R_2$ network at the output. Once again, the establishment of an ideal junction voltage v_f equal to the signal voltage without the error voltage $i_f r_B$ is desired.

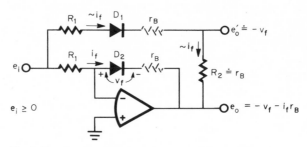

Fig. 3.23 A signal voltage equaling the error created by the bulk resistance r_B can be subtracted from the logarithmic amplifier output as shown.

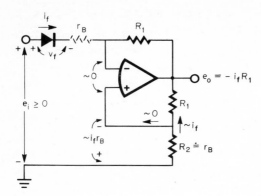

Fig. 3.24 A compensating signal voltage equaling the error voltage created by the bulk resistance r_B can be added to the input circuit of an antilog amplifier.

Without the compensation network the input signal e_i would support both v_f and $i_f r_B$, and so compensation is achieved by adding a correction signal to e_i. This signal is developed on R_2 by making it equal to r_B and by supplying it with a current equal to i_f. For $R_2 \ll R_1$, the current supplied to R_2 from e_o is very nearly at the desired level. By using the resulting correction signal to drive the noninverting amplifier input, the signal is essentially added to e_i. Summing the voltages around the input loop demonstrates that the ideal junction voltage v_f will now equal the signal e_i. Although the compensation network does supply positive feedback, it will always be less than the negative feedback for $R_2 < r_B + r_d$, where r_d is the dynamic resistance of the diode.

3.3.2 Improving frequency response   The most common logarithmic element used in logarithmic amplifiers is a transistor connected as shown in Fig. 3.25a. With this configuration the truest exponential relationship is attained, but the transistor adds gain to the feedback loop. The added voltage gain is approximately $R_1/(R_2 + r_e)$, where r_e is the dynamic emitter resistance of the transistor. While this gain is of some use in reducing low-frequency error, it also degrades frequency stability. Additional phase compensation is then required, as provided by the decoupling network $R_2 C$ described in Sec. 1.2.1. The phase compensation values are chosen for the highest operating current level since the transistor r_e is then lowest and gain is highest. At lower current levels this phase compensation is excessive and imposes an unnecessary frequency response limit.

This limitation can be eased by using a nonlinear phase compensation, as in Fig. 3.25b. In this circuit a nonlinear compensation resistance is formed, using a diode as one leg of a T network. Since the dynamic resistance of the diode is signal-dependent, the signal varies the phase compensation. At low signal levels the currents in the R_2' resistors are small, making the diode current low. The low diode current results in a high diode resistance that only slightly shunts the T. At higher signal levels the diode produces an increasing shunt effect to lower the gain around the feedback loop. In effect the equivalent resistance of the T is increased by this shunting, and this resistance increase counteracts the decrease in dynamic emitter resistance produced by the higher signal current. The result is a less signal-dependent gain around the feedback loop and less need for phase compensation capacitance.

3.4 Multiplier/Dividers

The analog multiplier is a highly versatile component in analog instrumentation. A variety of applications in this book make use of multipliers, including the voltage-controlled filters of the next section, the modulators and voltage-controlled oscillators described in Chapter 5, and one of the AGC circuits in Sec. 6.4. Operational amplifiers can be used to form multiplier/dividers in many ways.[1] Described in this section are circuits which make use of operational amplifier high-gain feedback to achieve multiplication and division with voltage-controlled resistors and with transistor logarithmic properties.

3.4.1 Voltage-controlled resistor techniques
A voltage-controlled resistor provides a very straightforward approach to multiplication and division. Ohm's law relates the current in a resistor to the quotient of the impressed signal voltage and the resistance. If the resistance is linearly

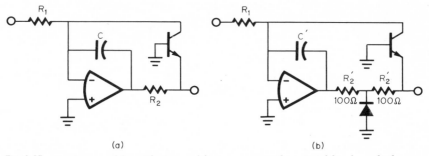

(a)   (b)

Fig. 3.25 For improved logarithmic amplifier response at low signal levels, a diode provides a signal-dependent phase compensation resistance.

proportional to a control voltage, the resistor current will be related to the quotient of the signal voltage and the control voltage. Alternatively, if the resistance is inversely proportional to a control voltage, the resistor current is proportional to the product of the two voltages. Two elements used for voltage-controlled resistors below are photoresistors and FETs.

To achieve precise voltage-controlled resistance, the resistance elements are connected in a basic gain-setting feedback loop,[2] as shown in Fig. 3.26 with photoresistors. Here the feedback loop around A_2 establishes the resistances of two photoresistors as dictated by two control voltages. To achieve this the feedback maintains near-zero input voltage and current for A_2, as shown. Under these conditions the voltage on photoresistor R_1 of A_2 equals Z, and the current through the resistor is fixed at Y/R_2. With both the voltage and current of the resistor fixed, its resistance is also fixed and will be $R_1 = R_2Z/Y$. If the other photoresistor is matched to this one, the inverter formed with A_1 will have a voltage-controlled summing resistor which is proportional to one voltage and inversely proportional to another voltage. The voltage on this summing resistor is controlled by yet another voltage X to result in an output voltage of $e_o = -XY/Z$. From this transfer function both multiplication and division can be attained.

Circuit performance is primarily dependent upon the characteristics of the photoresistors. Accuracy is essentially determined by the degree of match and tracking of these resistors. As long as they track, the photoresistors can exhibit nonlinear resistance versus light density, and the

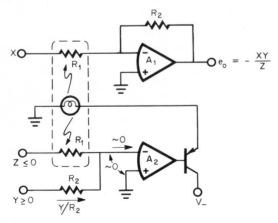

Fig. 3.26 A gain-setting feedback loop with photoresistors provides a multiplier/divider circuit.

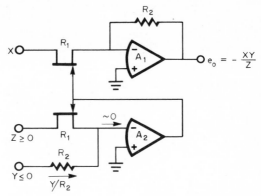

Fig. 3.27 FETs in place of the photoresistors of the previous circuit provide a speed improvement.

feedback loop will linearize the dependence upon the control voltages. Tracking error between the photoresistors typically produces a linearity error of 1 or 2 percent with this circuit. Input signal polarity requirements limit operation to two quadrants. For negative feedback to result, Z must be negative, and then Y must be positive to accept the current from Z. Input offset voltages on the operational amplifiers appear as signal offsets and result in signal feedthrough even when X or Y is zero. Frequency response is limited by the response of the lamp.

For a higher-speed multipler/divider, the photoresistors of the above circuit can be replaced with FETs. This results in the circuit of Fig. 3.27, with the FETs driven directly from the feedback amplifier, and avoids the slow response of the lamp above. Below pinchoff the FETs are essentially voltage-controlled resistors which are controlled by their gate voltage. While this control has a temperature-sensitive, nonlinear resistance-versus-voltage characteristic, the feedback amplifier A_2 makes the resistance stable and linearly related to the Z and Y signals as before. However, the resistances of the FETs are sensitive to their drain-source signal voltages as well as to the gate-source control voltage. This signal dependence results as the signal voltages drive the FETs toward their pinchoff voltages. Thus, unequal X and Z signals result in mismatched FET resistances and an associated multiplier/divider linearity error of several percent. The circuit permits only two-quadrant operation due to signal polarity restrictions for negative feedback, as described for the last circuit.

Significant improvement in the errors of the preceding FET multiplier can be made by reducing the signal voltages on the FETs and by using local feedback around the FETs. These modifications are shown in Fig. 3.28, where the input voltage dividers provide the signal reduction and the resistors labeled R_3 supply feedback. With the reduced signals the FET drain-source voltages will remain well below pinchoff, giving low signal-induced resistance changes. However, the reduced signals are now far more sensitive to the input errors of the operational amplifiers, such as input offset voltage and input bias current. The FET resistance signal sensitivity is decreased even further by feeding part of the drain or input signal back to the gate,[6] as shown. With the improvements of this circuit, the FET multiplier/divider will have a nonlinearity of 1 to 2 percent.

3.4.2 A logarithmic amplifier technique   Multiplication and division can also be achieved through the addition and subtraction of the logarithms of signals. Such operation is performed by the connection of logarithmic amplifiers and an antilog amplifier shown in Fig. 3.29. In this circuit the amplifiers A_1, A_2, and A_3 form logarithmic amplifiers with their associated transistors following the basic logarithmic amplifier from Fig. 3.25a. As before, phase compensation is added in the logarithmic amplifier feedback loops to compensate for the effects of the gains added

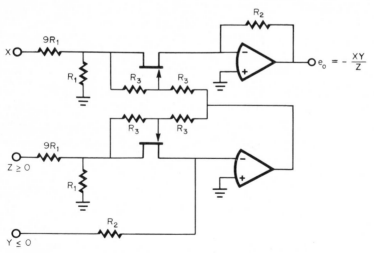

Fig. 3.28 Reduced input signals and local feedback around the FETs result in considerable reduction in the linearity error of the preceding FET multiplier/divider.

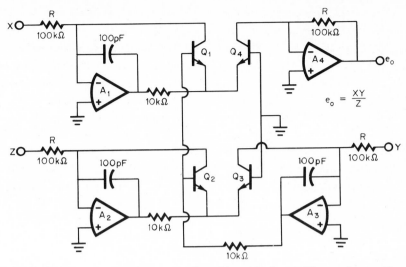

Fig. 3.29 By appropriate interconnection of logarithmic amplifiers and an antilog amplifier, multiplication and division can be provided by logarithmic techniques.

by the transistors. Amplifier A_4 forms an antilog amplifier with Q_4 to develop the output voltage $e_o = i_{c4}R$, where i_{c4} is set by the emitter-base voltage impressed upon Q_4.

The relationship between i_{c4} and v_{be4} can be expressed, using the junction equation,[1] as

$$i_{c4} = a\,i_{e4} \doteq aI_s e^{qv_{be4}/KT}$$

From this expression it is seen that the current which creates the output voltage bears an antilog relationship to v_{be4}. This emitter-base voltage equals the sum and difference of the output voltages of the three logarithmic amplifiers, as expressed by

$$v_{be4} = v_{be1} - v_{be2} + v_{be3}$$

Each of the logarithmic amplifier output voltages is developed by the current from an input signal, as can be expressed using the junction equation to give

$$v_{be4} = \frac{KT}{q} \ln \frac{i_{c1}i_{c3}}{aI_s i_{c2}} = \frac{KT}{q} \ln \frac{XY}{a\,I_s R\,Z}$$

where it is assumed that the transistors have the same a and the I_s. By substituting the last expression in the previous one for i_{c4}, the output voltage of the multiplier/divider can be expressed as $e_o = XY/Z$.

Although this logarithmic technique is more complicated than the voltage-controlled resistor approach, the performance of this circuit is also better. A nonlinearity error of less than 1 percent can be attained by using well-matched transistors. Operation is, however, restricted to one quadrant for both multiplication and division, since the input currents to the transistor collectors can be of only one polarity. Input offset errors are developed in this circuit by the input offset voltages and the input bias currents of the operational amplifiers. Moderate response speed can be achieved, as can be predicted from logarithmic amplifier response.

3.5 Active Filters

Precisely tailored frequency response characteristics can be produced with operational amplifiers in active filter circuits. The general circuit techniques employed have been extensively treated.[1] Described in this section are some convenient techniques for realizing tunable filters and for removing dc offset in both low-pass and high-pass filtering.

3.5.1 Tunable active filters

Tunable or variable active filters pose special circuit problems due to the interaction of adjustments and the need for manual tuning. In this section circuits are described which permit phase variation without disturbing gain magnitude and which permit electronic tuning. For phase adjustment a circuit is required which has variable phase shift but a gain magnitude that is constant with frequency. This performance is well approximated by the circuits of Fig. 3.30, where phase shifts are controlled by potentiometers. To visualize the operation of the circuits, the extremes of the control resistance can be considered. When $R = 0$ in Fig. 3.30a, the input signal is connected directly to the noninverting amplifier input, resulting in a voltage-follower configuration. This configuration produces zero phase shift, but as R becomes large compared to the impedance of C, the circuit functions like an inverter, with a phase shift of $-180°$. Analogous observations can be made for the other circuit.

The gain responses of the a and b configurations are

$$A_a = \frac{1 - j\omega RC}{1 + j\omega RC} \quad \text{and} \quad A_b = -A_a$$

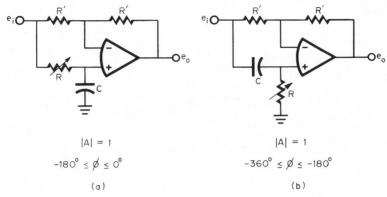

$$|A| = 1$$

$$-180° \leq \phi \leq 0°$$

$$(a)$$

$$|A| = 1$$

$$-360° \leq \phi \leq -180°$$

$$(b)$$

Fig. 3.30 Variable phase shift with constant gain magnitude is produced by the two circuits shown.

Note that the responses above have a coincident response pole and zero, making the gain magnitude constant with frequency. However, both the pole and zero do create phase shift as controlled by the variable resistors. The phase shifts developed by the two circuits are

$$\phi_a = -2 \tan^{-1} \omega RC \qquad \text{and} \qquad \phi_b = -180° + \phi_a$$

The accuracy of these expressions and the stabilities of the resulting phase shifts are essentially determined by R and C. Once again, errors due to the operational amplifiers are reduced to negligible levels when the feedback-loop gain is high. At higher frequencies the operational amplifiers introduce additional phase shift and produce a gain magnitude decrease. Any desired amount of phase shift can be achieved by using one or the other of the two circuits, as indicated by the ranges of phase shift shown with the circuits.

In general, tunable active filters are adjusted manually, using potentiometers. Electronic tuning provides opportunities for automatic adjustment, and such tuning can be accomplished with analog multipliers in active filters. By multiplying the signal voltage impressed on a resistor, the resulting current is increased as though the resistance were divided by the same factor. In other words, a change in the effective time constant can be achieved. Analogous observations can be made considering signal division or control of the signal voltages applied to capacitors. With this ability to control time constants, very rapid adjustments to filter characteristics can be made by varying the control voltages on multiplier and divider inputs. This technique is described below for basic low-pass, high-pass, and bandpass active filter stages.

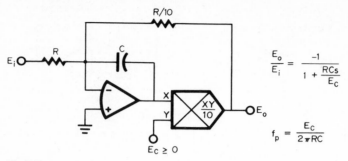

$$\frac{E_o}{E_i} = \frac{-1}{1 + \dfrac{RCs}{E_C}}$$

$$f_p = \frac{E_C}{2\pi RC}$$

Fig. 3.31 A multiplier feedback around an integrator produces a voltage-tunable low-pass filter.

Using these basic stages, or just their tuning techniques, more complex tunable filters can be formed.

The voltage-tuned low-pass filter of Fig. 3.31 is simply an integrator with a second feedback loop through a multiplier. Since this feedback is multiplied by a control voltage E_C, the response time constant is decreased. The result is a response time constant controlled by E_C, as illustrated by the expression

$$E_o = \frac{-E_i}{1 + RCs/E_C}$$

From this expression it is seen that low-frequency gain is unaffected by the control, and that the response pole frequency is directly related to the control voltage by $f_p = E_C/2\pi RC$. The accuracy of this control and of the overall response of the filter is directly affected by the multiplier gain error, nonlinearity, offset, and frequency response. Tuning range is limited by these multiplier errors and the resulting sensitivity to the control signal. Similarly, the speed of tuning is limited by the multiplier response.

For a voltage-tunable high-pass filter, the inverse of the above circuit can be used. It is shown in Fig. 3.32 and consists of a differentiator-type circuit with a divider in the input signal path. Since the signal applied to the capacitor is divided, the effective capacitance is multiplied, and the differentiator gain term is decreased. The overall output signal is

$$E_o = -\left(1 + \frac{RCs}{E_C}\right)E_i \qquad \text{for } R_L \ll R$$

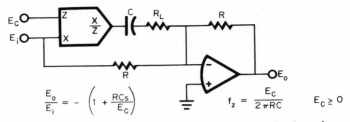

$$\frac{E_o}{E_i} = -\left(1 + \frac{RCs}{E_c}\right) \qquad f_z = \frac{E_c}{2\pi RC} \qquad E_c \geq 0$$

Fig. 3.32 A divider in the input of a differentiator-type circuit produces a voltage-tunable high-pass filter.

As desired, this response function is the inverse of that developed before with the low-pass filter. Once again, the response break frequency is $f_z = E_c/2\pi RC$, except that the frequency is now that of a response zero. Tuning accuracy and response are determined by the divider, and the various divider error and response limitations are added to the filter characteristics.

A multiplier can also be used to control the center frequency f_o of a bandpass filter. In Fig. 3.33 this control is added to the versatile, state-variable active filter configuration. In addition to providing response control, the multiplier removes the inverting amplifier common to such

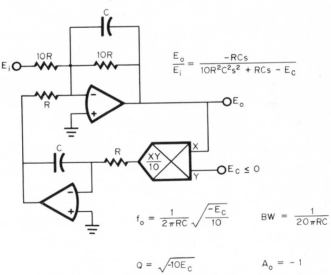

$$\frac{E_o}{E_i} = \frac{-RCs}{10R^2C^2s^2 + RCs - E_c}$$

$$f_o = \frac{1}{2\pi RC}\sqrt{\frac{-E_c}{10}} \qquad BW = \frac{1}{20\pi RC}$$

$$Q = \sqrt{-10E_c} \qquad A_o = -1$$

Fig. 3.33 A multiplier in the feedback loop of this bandpass filter provides voltage control of center frequency.

filters by providing negative gain in the loop. The filter is composed of two integrators in a feedback loop with a multiplier for voltage control of the feedback. In this way the filter response is controlled by E_C as expressed in

$$E_o = \frac{-RCsE_i}{10R^2C^2s^2 + RCs - E_C}$$

The control voltage does not move the midband gain from its level of −1. Control over center frequency is nonlinear due to the multiplier action and is expressed in

$$f_o = \frac{1}{2\pi RC}\sqrt{\frac{-E_C}{10}}$$

Bandwidth is not affected by the control voltage and remains at $BW = 1/20\pi RC$. Since bandwidth is constant, the response Q will vary along with f_o and will be $Q = \sqrt{-10\,E_C}$. Component accuracies directly affect the response precision, as with the basic state-variable filter. By adding the multiplier in the feedback loop of the filter, the errors and response limitations of the multiplier are added to the filter response.

3.5.2 Removing dc offset
With active filters, dc offset voltages are added to signals by the dc errors of amplifiers and by bias level shifts. Such offsets can often be removed from ac signals or avoided on dc signals by using simple filtering techniques. For ac signals a coupling capacitor provides a straightforward means of removing dc offsets; however, the resulting low-frequency cutoff is related to the signal source resistance. When the source resistance is quite low, such as that presented by an operational amplifier output, a very large capacitance is needed to establish the cutoff at a low frequency. Larger capacitors tend to have higher leakage, as well as greater physical size, so that the straightforward coupling capacitor can impose other limitations. To reduce the level of capacitance required, a resistor can be added in series with the signal source, but this directly increases output resistance.

Alternatively, integrator feedback can be used to remove dc offset from ac signals without a coupling capacitor. Two such feedback circuits are shown in Fig. 3.34. When the signal source can be floated, the circuit of Fig. 3.34a can be used. In this case the signal source receives a dc shift from an integrator which compares the output dc level with

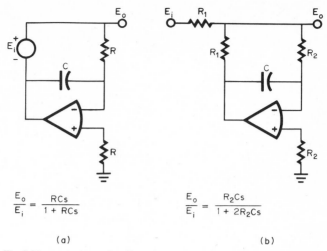

$$\frac{E_o}{E_i} = \frac{RCs}{1 + RCs}$$

$$\frac{E_o}{E_i} = \frac{R_2Cs}{1 + 2R_2Cs}$$

(a)   (b)

Fig. 3.34   To remove dc offsets from ac signals without using a large coupling capacitor, a feedback integrator can be connected to supply an offset compensating voltage.

ground. When the output reaches a zero dc level, the integration stops and the integrator holds the negative side of the source at an offset compensating dc level. In this way the output offset developed by the source is reduced by a factor equal to the dc open-loop gain of the operational amplifier. An additional output offset voltage component is, however, added by the input offset voltage of the operational amplifier. With this circuit the low-frequency cutoff is set by the integrator at $f_c = 1/2\pi RC$. The integrator resistor R can be in the megohm range, as compared with the few ohms often presented to a coupling capacitor, and so a very much smaller capacitance can be used to achieve a given cutoff frequency.

Where the signal source cannot be floated, the integrator feedback circuit of Fig. 3.34b can be applied. Once again the integrator feeds a signal back to the input to reduce the dc level of the output to zero. In this case the feedback signal is summed with the input signal on the resistors labeled R_1 to remove dc offset. This reduces the offset from the signal source by the open-loop gain of the amplifier, but adds an offset equal to the input offset voltage of the amplifier. The summing resistors labeled R_1 divide the signal transmitted by a factor of 2 and increase the circuit output resistance. However, the increase in output resistance is far less than would be required to achieve a low cutoff fre-

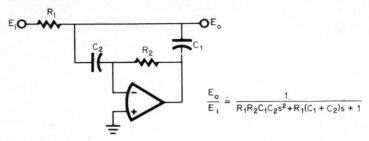

$$\frac{E_o}{E_i} \doteq \frac{1}{R_1 R_2 C_1 C_2 s^2 + R_1 (C_1 + C_2) s + 1}$$

Fig. 3.35 With this two-pole, low-pass filter, no amplifier dc offset is added to the signal since the amplifier is ac-coupled.

quency with a small coupling capacitor. Now the cutoff frequency is a function of the integrator resistor and will be $f_c = 1/4\pi R_2 C$.

In low-pass filter circuits dc offsets are even more significant since they cannot be distinguished from dc signals. In some cases the dc offset of low-pass filters can be removed by using the filter structure of Fig. 3.35. This circuit is a two-pole, low-pass active filter which has no dc offset introduced by the operational amplifier. One response pole is developed by the simple RC filter formed with R_1 and C_1, and the second pole is produced by the ac amplifier formed by R_2, C_2, and the operational amplifier. To produce the second pole, the ac amplifier drives C_1 with an out-of-phase signal. Since the operational amplifier is ac-coupled at its input and output, it cannot add any dc offset to the signal. The filter is limited to a dc gain of unity and has an output resistance equal to R_1. To minimize R_1, and the resulting output resistance, this resistor should be used for the higher-frequency pole, and C_1 should be maximized. A further limitation of this filter is a response zero that results at high frequencies where the operational amplifier gain becomes insufficient.

REFERENCES

1. G. E. Tobey, J. G. Graeme, and L. P. Huelsman, *Operational Amplifiers: Design and Applications*, McGraw-Hill Book Company, New York, 1971.
2. G. A. Korn and T. M. Korn, *Electronic Analog and Hybrid Computers*, McGraw-Hill Book Company, New York, 1964.
3. *Handbook of Operational Amplifier Applications*, Burr-Brown Research Corporation, Tucson, Ariz., 1964.
4. W. Pierce, Constant Current Sources Gain New Capabilities, *EDN*, Sept. 1, 1970.
5. S. Franco, Op Amp Log Circuit Eliminates Bulk Resistance Effects, *Electronics*, June 9, 1969.
6. T. Mollinga, The FET as a Voltage Controlled Resistor, *EEE*, January 1970.

4

SIGNAL PROCESSORS

Operational amplifier signal processors analyze, route, rectify, sample, and convert electrical signals. In this chapter circuits which perform these functions are described, including comparators, multiplexers, clamps, absolute-value circuits, sample-hold circuits, and rms converters. Extensions of the basic comparator connections which provide modified functions and circuit simplifications are described. Improved precisions are provided for multiplexing and clamping operations by operational amplifiers, as is shown. Similar improvements are provided for full-wave rectification when performed by absolute-value circuits, which are described in numerous forms. Accuracy and speed improvements for sample-hold circuits are outlined as developed with various circuit modifications. Precise measurement of signal magnitude is provided by the computing and thermal rms-to-dc converters.

4.1 Comparators

Analog comparators are common elements in signal processing, used for discriminating between signal levels. An operational amplifier without feedback serves as a basic comparator by comparing the signals at its inputs. When an operational amplifier is used, hysteresis can be added by simply supplying positive feedback. The phase compensation of the amplifier should generally be removed in these applications to improve switching speed. At low signal levels the switching speed is limited by the gain-bandwidth product of the amplifier, and at high signal levels it is limited by the slewing rate. Interconnected amplifiers can form a window comparator for detection of signals within a given band. Each of these comparator connections is described in the first reference.[1] Additional circuits described here include specialized comparators and simplified window comparators.

4.1.1 Specialized comparators
The common operational amplifier comparator connection can be modified to provide hysteresis without altering the trip point,[2] and it can be modified to provide detection of the slope rather than the amplitude of a signal. For adjustable hysteresis and a constant trip point, a unipolar hysteresis feedback is connected, as shown in Fig. 4.1. In this circuit the signal diode interrupts one polarity of positive feedback supplied through R_2. Then, hysteresis is developed for only one comparator state, and one trip point remains at the original level set by the reference E_R. The second trip point added by hysteresis is removed from the original trip point by a hysteresis of

$$\Delta V = \frac{R_1(V_Z - E_R)}{R_1 + R_2} \qquad \text{where } V_Z > E_R$$

By varying R_2, the hysterisis can be adjusted without disturbing the trip point at E_R.

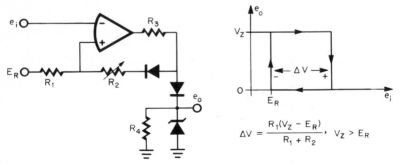

Fig. 4.1   A comparator connection with a signal diode in the hysteresis feedback loop provides adjustable hysteresis along with a constant trip point.

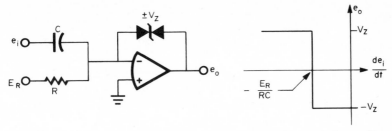

Fig. 4.2 A slope comparator is formed by replacing a summing resistor with a capacitor.

Other performance characteristics are similar to the common operational amplifier comparator circuit. Both trip-point accuracies are limited by the input offset voltage, input bias current, and finite gain of the operational amplifier. The output signal can be taken either directly from the amplifier output or from the zener diode as shown. With the latter connection the output signal alternates between zero and V_Z, as might be desirable for interface with digital logic circuits. However, in the zero volt state this output cannot sink current. Resistor R_3 is added in series with the signal diode to limit the current drain through the diode, and resistor R_4 is connected to provide a discharge path for the diode capacitance.

Another variation on the basic comparator connection permits comparison of the rate of change of a signal against a constant. Actually, any two signal characteristics can be compared by an operational amplifier if the difference between the characteristics can be made to develop a nonzero voltage between the amplifier inputs or to force a nonzero current into either input. When either of these nonzero conditions is established, over and above the input offset voltage or input bias current of the amplifier, the amplifier output is driven to one of its voltage limits. In this way the slope comparator of Fig. 4.2 switches its output state when the rate of change of the input signal exceeds a constant. The current developed in C by e_i is compared with that produced in R by E_R, and switching occurs when the combined current changes polarity. Since this change occurs at the zero current crossing, the trip point is

$$\frac{de_i}{dt} = -\frac{E_R}{RC}$$

Trip-point accuracy is limited directly by the input offset voltage, input bias current, and finite gain of the amplifier. The feedback limiter shown is needed to absorb the current from E_R following transitions.

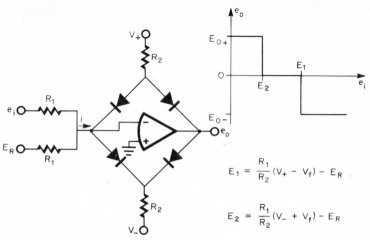

Fig. 4.3 With a diode bridge for feedback, an operational amplifier performs as a window comparator.

Without such a limiter the current from E_R would charge C and, thereby, offset the amplifier input from zero. With two input capacitors the circuit can provide comparison of the slopes of two signals in an analogous manner.

4.1.2 Simplified window comparators

A window comparator indicates when a signal is within a certain band, and some comparators of this type also indicate whether a signal outside the band is above or below the band. These comparators are often formed with two operational amplifiers and two voltage references. Described in this section are a single amplifier configuration and a circuit not requiring external references. The single amplifier form is that of Fig. 4.3, which consists of an operational amplifier enclosed in a feedback diode bridge. With low input current i, the diode bridge provides low impedance feedback to hold the amplifier output near zero. As signals raise the input current, the output voltage change is initially only that created by changing diode currents. However, when the comparator input current exceeds that which can be supplied by the diode bias, the operational amplifier input is driven away from ground level. This drives the amplifier output into saturation, E_{O+} or E_{O-}, and reverse-biases the feedback diodes. The current levels at which this happens are the positive and negative trip points of the comparator. Expressing these trip points in terms of the input signals, the comparator operation is described by threshold voltages of

$$E_1 = \frac{R_1}{R_2}(V_+ - V_f) - E_R \quad \text{and} \quad E_2 = \frac{R_1}{R_2}(V_- + V_f) - E_R$$

where V_f is the forward voltage drop of a diode.

Both the amplifier and the diode bridge limit the accuracy of the comparator. As commonly encountered, the amplifier input offset voltage, input bias current, and finite gain add errors directly to the trip points. Further trip-point error is introduced by the variations in diode drops, but the trip point can be made independent of the diodes by using current sources to bias the bridge. The diode bridge also creates a midband error, as the input current i unbalances the diode voltages. As this current varies, the voltage unbalance produces an output offset voltage of V_f to $- V_f$. Output switching is rate-limited to the slewing rate limit of the operational amplifier.

For a high-resolution window comparator with trip points of a few millivolts, the reference voltages can sometimes be replaced by offset voltages developed in operational amplifiers. The input offset voltage of an operational amplifier directly shifts its comparator switching threshold. As a result, the offset voltage null control of an operational amplifier can be used to set a trip point at a level of a few millivolts. Using this approach, the differential input window comparator of Fig. 4.4 produces the transfer characteristic shown. The trip points of this com-

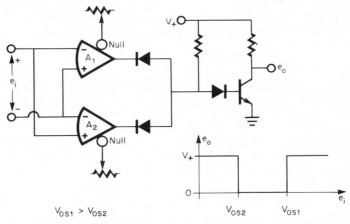

Fig. 4.4 Low-voltage trip points for this window comparator are set by the offset null controls of the operational amplifiers.

parator will generally be temperature-sensitive due to the effect of the null control on offset voltage drift described in Sec. 1.1.2. Where this represents a problem, a temperature-compensated null control can be applied, as described in Sec. 1.1.3. Included in the window comparator of Fig. 4.4 is a gating circuit that decodes the output signals from the two amplifiers. Only when both amplifier outputs are at their high-voltage state does the gated output reside at zero voltage.

4.2   Multiplexers and Clamps

Multiplexing and clamping are common operations in instrumentation that can be significantly improved by using operational amplifiers. Multiplexing greatly reduces the number of signal processing channels or signal lines required in a system by permitting time-sharing. Clamping serves a number of instrumentation needs, including signal clipping, signal squaring, and overload protection. For all these functions, errors can be greatly reduced by the high-gain feedback of operational amplifiers.

4.2.1   Multiplexing amplifiers   In order to multiplex various signals to a point, there must be switches in the signal paths. For high-speed multiplexing, these switches are generally solid-state devices which can introduce errors due to offset voltage, ON resistance, leakage current, and capacitance. Most of the switch errors can be removed or compensated by using operational amplifiers to form multiplexing amplifiers. Described here are a multiplexer with an error-compensating feedback network, a multiplexer with switch errors reduced by the gain of an amplifier, and an application of the latter in a multiplying digital-to-analog converter.

A common multiplexing amplifier connection incorporates switchable summing resistors on an inverting amplifier, as shown in Fig. 4.5. The operational amplifier summing junction holds the sources of the MOS-FET switches at ground potential. As a result, the signal swing does not alter the gate-source turn-on voltage as it does when the switches supply a resistor load. Without signal swing on the gate-source biases, the ON resistances of the switches are not modulated by the signal, and the gate-source capacitances do not shunt the signal. However, even a signal-independent switch resistance introduces a gain error through the multiplexer. This error can be corrected at one temperature by a

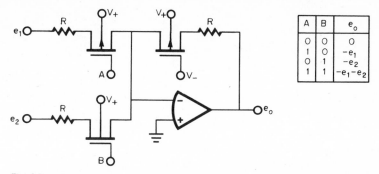

A	B	e_o
0	0	0
1	0	$-e_1$
0	1	$-e_2$
1	1	$-e_1-e_2$

Fig. 4.5   The MOSFET in the feedback path of a multiplexing amplifier compensates for the ON resistance errors of the switches.

decrease in the associated summing resistor, but the drift of the ON resistance creates a gain drift. To compensate both the gain error and its drift, a resistance matching the ON resistance can be added to the amplifier feedback resistance. This is accomplished in Fig. 4.5 by connecting a matching MOSFET in the feedback loop. With matching MOSFETs the ON resistance error can be reduced an order of magnitude. Then the remaining multiplexer errors will be the switch leakage in the OFF state and the normal errors of an operational amplifier inverter.

It is desirable to amplify signals from remote sources before sending them over long wires to a multiplexer. The preamplifiers used for this can be switched so that they also perform the multiplexing function. In addition, this combination of switch and amplifier forms a precision analog gate with accuracy that approaches that of an ideal multiplex switch. This technique is applied in Fig. 4.6 with complementary MOSFET switches that control the feedback of an inverting amplifier.[3] Since the switches are in the feedback loop, the effects of their ON resistances are divided by the feedback-loop gain. If bipolar transistor switches were used, the effects of their offset voltages would also be reduced by the loop gain.

The operation of the precision analog gate shown is controlled by the MOSFET drive signal A. When this signal is at its high level, Q_2 is OFF, and Q_1 is ON to conduct the feedback current through R_2. This results in an inverting amplifier configuration with an output of $e = -R_2e_i/R_1$ and the normal errors of an inverting amplifier. Added to these errors is the effect of the ON resistance of Q_1. The voltage drop on this resistance, ir_{ON}, is supported by a reduced summing junction voltage of

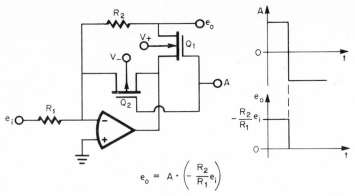

$$e_o = A \cdot \left(-\frac{R_2}{R_1} e_i \right)$$

Fig. 4.6   By connecting switches in the feedback loop of an operational amplifier, their ON resistance errors are reduced by the feedback-loop gain.

$i r_{ON}/A$, where A is the open-loop gain of the amplifier. From this summing junction voltage an amplifier output error voltage of $i r_{ON}/A_L$ results, where A_L is the loop gain. Then, the effect of r_{ON} is reduced by A_L. In the opposite switch state Q_1 is OFF, and Q_2 is ON to conduct the feedback current. Now the output terminal simply sees the virtual ground of the operational amplifier input through R_2. This results in $e_o \doteq 0$ within the error produced by the input offset voltage of the amplifier. If the output terminals of a number of these precision gates are connected, they will establish an amplified output voltage set by whichever gate is ON.

The above analog gate can also provide high precision to another gating function in a multiplying digital-to-analog converter (MDAC). Such a converter permits digital control of the gain by which a signal is amplified, and a simple modification to the specific MDAC to be described converts it to a multiplexer. A partial circuit of the MDAC is shown in Fig. 4.7 with only one gating element or bit input. By adding other summing networks and gates in parallel with the set shown, the MDAC is formed. The summing networks would be binary-weighted to provide conversion from a binary digital code to an analog gain. Alternately, the added summing networks can be connected to separate signals to achieve multiplexing.

Each analog gate circuit either shunts or passes the signal that is on its associated summing resistor network by switching the R_2 element into and out of the feedback loop of the gate amplifier. As will be described, this alternately opens and closes a shunt switch around one element of a T summing network. In both switch states, one end of R_2 is connected to a virtual ground at the input of the gate amplifier. When Q_1 is OFF and Q_2 is ON, only the virtual ground end of R_2 is held by the analog gate, and R_2 forms the grounded leg of a T summing network. In this state the switch errors of the analog gate are the input offset voltage and input error signal of the gate amplifier. This offset voltage can be made quite small, and the error signal will equal the voltage developed by the feedback switch ON resistance divided by the open-loop gain of the gate amplifier.

In the opposite switch state, Q_2 is OFF and Q_1 is ON to short the common point of the T network to the output of the gate amplifier. This shunts current away from R_2 to the low impedance amplifier output with the same result as a shunt switch. With essentially no current in R_2, the common point of the T network is held at the virtual ground of the gate amplifier, and no signal is passed to the output amplifier. Switch errors in this mode are the input offset voltage, input bias current, and input error signal of the gate amplifier. Once again, the input error signal is that developed by signal current with the switch ON resistance divided by the open-loop gain of the gate amplifier. In both switch modes the offset and ON resistance errors will be very small compared to the errors of simple series or shunt switches, within the bandwidth limitations of the gate amplifier.

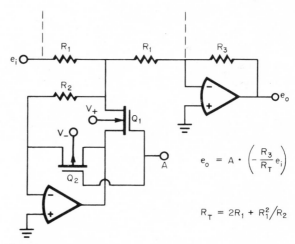

$$e_o = A \cdot \left(-\frac{R_3}{R_T} e_i \right)$$

$$R_T = 2R_1 + R_1^2/R_2$$

Fig. 4.7   The analog gate circuit of Fig. 4.6 provides precision switching of summing networks for MDAC or multiplexer operation.

4.2.2 Clamping amplifiers   Precision is also added to signal clamping by the high-gain feedback of operational amplifiers in a variety of circuit configurations.[1]   Additional clamping amplifier configurations described here are an adjustable clamp with a wide range of control and a connection which decouples the leakage current of the clamp element. One of the most common clamp elements is the zener diode, as it is simple and moderately precise.   These conveniences can be extended to an adjustable clamp with the circuit of Fig. 4.8.   With this circuit the portion of the input signal reaching the output terminal as e_o is controlled by the zener diode; and when the potentiometer reaches the conduction level of the zener diode, this diode conducts feedback current to limit e_o.

For a given zener diode the clamping level is established primarily by the gain-setting potentiometer.   It controls the gain by which the amplifier multiplies e_o to drive one end of the zener diode.   Clamping occurs when $e_i \doteq e_o = \pm K V_Z$, where V_Z is the zener voltage and K describes the potentiometer setting as shown.   Then, by simply varying the potentiometer, clamping can be achieved at almost any fraction of V_Z, and even millivolt-level clamping is possible.   However, low-level clamping is limited by the dc errors of the operational amplifier, including input offset voltage, input bias current, and gain error.   A further error is produced by the leakage current of the zener diode as a clamp level is approached.   Before this level is reached, output loading can degrade accuracy, since the output resistance at the output terminal will be $R_S \parallel KR$.   Beyond the clamp level the zener diode provides a low impedance return to the operational amplifier output and results in a low output impedance.

One of the error sources in the above clamping amplifier is the leakage current of the zener diode, and such an error is created by most clamp elements.   To avoid this error the leakage current of the clamp element can be decoupled from the input of the clamping amplifier,

$- K V_Z \leq e_o \leq K V_Z$

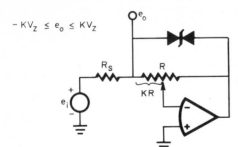

Fig. 4.8 An adjustable clamp is formed by using an operational amplifier to amplify the voltage impressed on a zener diode.

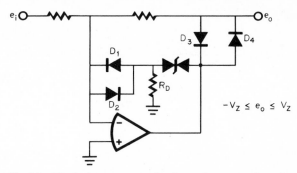

Fig. 4.9 The leakage current of a clamp element such as the zener diode can be decoupled from the input circuit without adding a diode voltage drop to the clamp levels.

as shown in Fig. 4.9. This circuit is a modified form of the simple bipolar clamp formed with a double-ended zener diode across the feedback element of an operational amplifier. As before, the large output voltage of the operational amplifier induces leakage current in the zener diode. However, in this case the leakage current is shunted to ground through the decoupling resistor R_D instead of being passed to the input. As long as the resulting voltage on R_D is small, diodes D_1 and D_2 will be OFF. The reverse-biased diodes isolate the amplifier input from the zener leakage current, except for the leakage current developed in D_1 and D_2. With the small voltage on these diodes, this new leakage current reaching the input is greatly reduced.

However, in the clamping states the forward voltage drops of these two diodes increase the clamp voltages at the amplifier output. Since the diode forward voltages are rather temperature-sensitive, the precision added by leakage current decoupling could be lost in thermal drift. To prevent this, diodes D_3 and D_4 are added to remove the diode drops from the clamp levels at the output terminal. If the forward voltages of D_3 and D_4 approximately match those of D_1 and D_2, the diode voltage errors will be nearly canceled, and clamping will occur at $e_o = \pm V_Z$. Then, the principal error sources will be those of the inverting amplifier. High-frequency operation is limited further by the addition of D_3 and D_4, since the operational amplifier output must now slew very rapidly in turning one diode OFF and the other ON at the zero crossing.

4.3 Absolute-value Circuits

By taking the absolute value of a signal it is converted from a bipolar to a unipolar form, as is often required for magnitude detection and other purposes. Magnitude detection in most average reading measurement instruments begins with an absolute-value conversion. With operational amplifiers various absolute-value or precision rectifier circuits can be formed to achieve the desired highly accurate full-wave rectification. In these circuits rectification is achieved without sacrificing a significant portion of the signal to forward-bias rectifying diodes. By connecting these diodes in the feedback loop of an operational amplifier, this signal loss is reduced by the high gain of the amplifier. With this connection the feedback drives the diodes into and out of conduction in response to very small signal changes, permitting rectification of millivolt-level signals. Numerous circuits providing this precision rectification are described in this section, including high-precision configurations, single amplifier connections, and circuits for improving frequency response.

4.3.1 Precision absolute-value circuits
The more precise absolute-value circuits are formed with two operational amplifiers connected so that the polarity of the net circuit gain switches when the polarity of the input signal reverses. In this way the polarity of the output signal is prevented from changing. This feature, coupled with equal gain magnitudes for input signals of either polarity, results in an absolute-value conversion. Along with the basic absolute-value circuit, three of its variations are presented in this section. These variations are a high input impedance connection, a configuration which accommodates input signal summation, and a simplified circuit requiring only two matched resistors.

Shown in Fig. 4.10 is the basic absolute-value circuit[1] which achieves rectification through feedback switching around A_1. For positive input

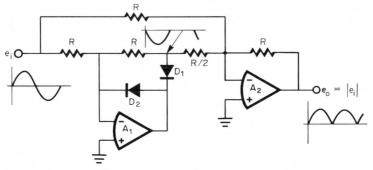

Fig. 4.10   The basic absolute-value circuit reverses the sign of the net circuit gain when a signal polarity change switches the feedback around A_1.

signals D_2 is OFF and D_1 conducts the feedback current to connect A_1 as an inverter. This inverted signal is summed with the original input signal by A_2. Since the inverted signal is impressed on a smaller summing resistor, it dominates the summation to produce a positive output signal. When the input signal is negative, D_1 is OFF and D_2 conducts the feedback current of A_1 to hold its input at its virtual ground. At this virtual ground all signal on one of the input paths is shunted, leaving only the original input signal as a signal to A_2. Inversion in this amplifier results in an output which is again positive. For the resistor weighting shown, the gain magnitude will be unity for both polarities of input signal. Higher gain is developed by increasing the feedback resistor of A_2. A separate polarity indication can be derived from the output state of A_1 or from a transistor connected with its emitter-base junction as one of the diodes.

A number of factors limit the performance of the above circuit, including resistor matching and the input offset voltages, input bias currents, slewing rate limits, and gains of operational amplifiers. Any deviation of resistor match from the ratios indicated will produce gain error, which in most cases will make the gain magnitudes different for the two polarities of input signal. The input offset voltage and input bias current of A_2 will create a simple output offset, but the analogous input errors of A_1 create different output errors for the two signal polarities. High-frequency rectification is limited by a deadband around zero resulting from the limited slewing rate and gain in A_1 and from diode capacitances. This limitation and some improvements on it are outlined in Sec. 4.3.3.

Most precision rectifier circuits have low input impedances set by input summing resistors, and an additional buffer amplifier may often be needed. However, the circuit shown in Fig. 4.11 avoids the need for an additional amplifier,[4] as it presents the high common-mode impedances of the amplifiers, instead of summing resistors, to the input terminal. This results in typical input resistances of 25 MΩ for bipolar transistor input amplifiers or 10^{12} Ω for FET input amplifiers. Except for this input connection and deletion of a summing resistor on A_2, this circuit is identical to the basic form described above. Full-wave rectification is again achieved by feedback switching around A_1 that turns on and off one of the signal paths to A_2. Positive-going signals develop a positive current for i_1, as shown, which turns D_1 to ON since neither D_2 nor the amplifier input can supply this current. With only D_1 at ON, A_1 is connected as a voltage follower and it forces point A to follow the input

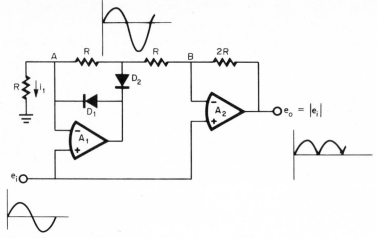

Fig. 4.11 A high input impedance absolute-value circuit is formed by some simple modifications to the basic circuit of Fig. 4.10.

signal e_i. Similarly, feedback around A_2 causes its two inputs to track so that point B also follows e_i. Then there is no voltage difference between points A and B, so that no signal current flows between the two points. As a result, no signal current is generated for the feedback path of A_2, and the output voltage will equal the input voltage. In summary, positive input signals cause A_1 to effectively turn off one signal path to A_2, and A_2 then performs as a voltage follower.

When the input signal is negative, both signal paths to A_2 are ON. The resulting negative current for i_1 turns D_1 to OFF and turns D_2 to ON to connect A_1 as a noninverting amplifier with a gain of $+2$. This amplifier presents a signal of $2e_i$ to A_2, and this latter amplifier appears as an inverting amplifier with a gain of -2 to this first signal path. To the other signal at its noninverting input A_2 appears as a noninverting amplifier with a gain of $+3$. Using superposition to combine the effects of the two signals, the output signal is found to be

$$e_o = -4e_i + 3e_i = -e_i \qquad \text{for } e_i < 0$$

Then, a change in input signal polarity changes the overall circuit gain from $+1$ to -1.

Both operating modes are limited by the errors described above for the basic circuit, with the exceptions that the modified circuit has higher input impedance, one less resistor to match, and a lower signal swing

capability. The reduced swing is imposed by A_1, which amplifies negative signals by a factor of 2 and will reach its output saturation limit at a
lower signal level than would A_2. Otherwise, the circuit errors are analogous to those described for the preceding circuit as created by resistor
mismatch and by the dc errors, slewing rate limits, and gains of the operational amplifiers.

Another limitation of the basic absolute-value circuit of Fig. 4.10 is that
signal summation at its input is more complex than normal. In that circuit the input signal is supplied to two points, but some simple modifications reduce this to one point at which signals can be summed easily.
The modified circuit is that of Fig. 4.12, and any number of signals can
be summed at its input by adding summing resistors. In this circuit the
feedback around A_1 switches to control the distribution of current i_1
between the two signal paths. As before, the switching is induced by a
change in input signal polarity, and it results in a change in net circuit
gain from $+1$ to -1. For a positive input voltage the current i_1 will
be positive and will forward-bias D_1 and reverse-bias D_2. This connects
A_1 as an inverter driving the inverting input resistor of A_2. With D_2
at OFF, the noninverting input of A_2 is held at the virtual ground of the
input of A_1. Then the combined circuit is connected as two cascaded in-
verters to give $e_o = e_i$ for positive input signals.

When the input signal is negative, D_1 is OFF and D_2 is ON. This connects A_1 as an inverter driving the noninverting input of A_2, and it is this
switch to the other input of A_2 that changes the sign of the circuit gain.

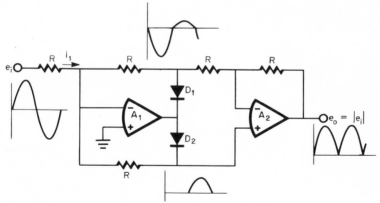

Fig. 4.12   An absolute-value circuit with input summing capability results from
modified feedback switching around A_1.

In this mode all i_1 does not flow in the new feedback resistor around A_1, as signal current continues to flow in the other signal path. Both signal paths lead to the same potential at the inputs of A_2, so that i_1 divides between the two paths as determined by their resistances. For the resistances shown, two-thirds of the current flows in the feedback resistor of A_1, establishing its output voltage at $-2e_i/3$. The remaining third of the current flows in the opposite path through the feedback resistor of A_2, developing a voltage on this resistor of $-e_i/3$. From the combination of these signal voltages, the output voltage becomes $e_o = -e_i$ for negative input signals. The accuracy and speed limitations of this circuit are analogous to those described for the basic circuit at the beginning of this section except that resistor matching is made easier because all resistors here have the same value.

A significant difficulty encountered in building any of the preceding absolute-value circuits for high accuracy is the need for accurate ratio matching of four or five resistors. This task is simplified with the circuit of Fig. 4.13.[5] While this circuit lacks the high input resistance of that of Fig. 4.11 and lacks the input summing capability of the last circuit, it requires only two matched resistors. If the resistors labeled R_1 are matched, the gain magnitudes presented to the two signal polarities will be matched. The value of R_2 is somewhat arbitrary, and it is chosen

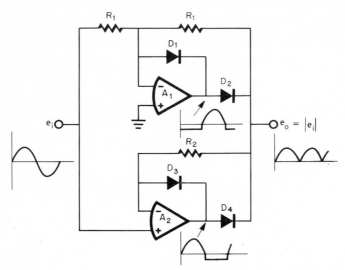

Fig. 4.13 An absolute-value circuit with only two matched resistors is formed by an inverter and a voltage follower with output gating that selects the positive output voltage.

to develop low error with the input bias current of A_2 in one mode but to present a moderate load to A_1 in the other mode.

Basically the circuit consists of an inverter and a voltage follower connected in parallel with diode gating to select the output which is positive. The gating diodes are enclosed in the amplifier feedback loops so that they introduce only small errors. When the input signal swings positive, the output of A_1 becomes negative and the output of A_2 becomes positive. These outputs reverse-bias D_2 and forward-bias D_4 to connect A_2 to the output terminal for a voltage-follower configuration. No signal current flows in R_2 in this state, as D_3 is reverse-biased by the positive output, and this makes the output voltage essentially equal to the input voltage. In this mode the output of A_1 is clamped by D_1 to prevent saturation and an associated switching delay.

For negative input signals A_1 develops the positive output voltage and is connected to the output terminal of the overall circuit. This results in an inverter circuit connection to give $e_o = -e_i$. In this state A_2 is clamped by D_3 to avoid saturation, and A_2 draws current through R_2 from A_1. Since a different amplifier is now connected to the output, the output offset created by dc amplifier errors will have changed with the switching. Switching speed and the associated response limit are determined by the slewing rate limits and gains of the amplifiers as detailed in Sec. 4.3.3.

4.3.2 Single amplifier configurations   Each of the absolute-value circuits of the previous section uses two operational amplifiers to develop precise full-wave rectification. For moderate accuracy requirements the circuits of this section provide absolute-value conversion with one amplifier. Once again, the circuits have some means of switching the net circuit gain from $+1$ to -1 when the input signal polarity reverses. One of the circuits switches from a voltage-follower configuration to an inverter configuration, and the last two circuits employ current rectifiers. The follower/inverter circuit is illustrated in Fig. 4.14. In this circuit the polarity of the input signal is detected by the zero-crossing switch formed with Q_3 and Q_4. This switch drives Q_1 to alter the effective circuit configuration.

For negative input signals Q_4 is at ON to turn Q_1 to ON and to ground the noninverting input of the amplifier. In this state the amplifier operates as an inverter, with Q_2 serving simply as a compensation resistance.

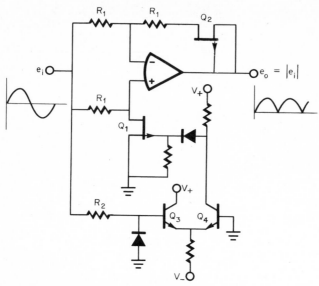

Fig. 4.14 A simple absolute-value circuit is formed with a zero-crossing switch which converts an inverter to a voltage follower.

This added FET behaves as a resistor as long as its current is below I_{DSS}, and the voltage developed on this resistance compensates for that lost on the switch Q_1. When the input signal becomes positive, both Q_4 and Q_1 are turned to OFF. Then the noninverting input is returned to the input signal through a resistor. Essentially no signal current flows through this resistor to the amplifier input, and so this input follows the input signal voltage. The inverting input of the amplifier will follow this signal on the noninverting input, so that no signal voltage is impressed on the R_1 summing resistor. As a result, no signal current is supplied to the R_1 feedback resistor, and the output follows the input signal. In other words, the circuit operates as a voltage follower.

Performance is primarily limited by the errors of switch Q_1, with smaller errors contributed by the operational amplifier. When at ON, this switch has an ON resistance error which is largely compensated by means of Q_2. The residual ON resistance error voltage has the effect of a gain error, since it is related to signal current. A compensating gain adjustment can be made by adjusting the value of the summing resistor. In the OFF state Q_1 creates a leakage current flow in its input resistor to develop an error voltage. A similar error is developed by the input bias current of the amplifier. All the errors developed by the amplifier are the same as those developed in the common inverter and voltage-follower circuits. Another source of error can be the extreme change in circuit input impedance as its state switches. This impedance changes from $(R_1/2) \parallel R_2$ to the common-mode input impedance of the amplifier, and this can result in a severe change in input loading effects.

Where the input signal is a current, the absolute-value conversion can be performed with the simpler circuit of Fig. 4.15. In this case the signal itself switches the connection to the amplifier. For a negative signal current D_1 is ON and the current flows in the feedback resistor to develop an output voltage of $e_o = - i_i R$. When the input signal current is positive, D_2 is turned to ON to pass the current to the noninverting input. Here it develops a voltage on the input resistor that is followed by the amplifier to give an output of $e_o = i_i R$. In summary, the signal current is gated by the diodes to that input which will result in a positive output.

Another absolute-value conversion technique makes use of the ease with which currents can be rectified. When the signal is a current instead of a voltage, no error is introduced by the forward-biasing voltages of rectifying junctions. To make use of this property, the controlled current source techniques of Sec. 3.2 can be used for voltage-to-current conversion. In this way the absolute-value circuit of Fig. 4.16 is formed, with Q_1 and Q_2 as current output transistors for opposite current polarities. The operational amplifier drives either Q_1 or Q_2 to whatever level is required to make the inverting input follow the noninverting input. Then a voltage equal to the input signal is established on R_1, and the associated current is supplied through either Q_1 or Q_2.

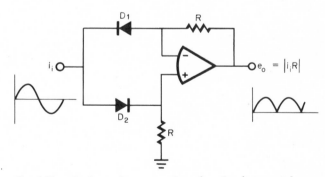

Fig. 4.15 Absolute-value conversion of a signal current is performed in this circuit by signal switching of input connections.

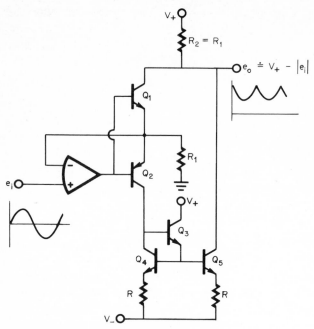

Fig. 4.16 Absolute-value conversion is achieved here by converting the signal to a current and rectifying the current.

If the input signal is positive, Q_1 supplies the current from its emitter, and the resulting collector current flows in the load resistor R_2 to develop the output voltage. In this state no current is supplied to the load resistor from Q_5 as this transistor has zero base bias when Q_2 is OFF. Then the load current is essentially the current in R_1 less the base current of Q_1. For $R_2 = R_1$, this results in $E_o \doteq e_i$. When the input signal is negative, Q_1 is OFF, and Q_2 supplies the current to R_1. Collector current from Q_2 flows in Q_4 to develop base bias for current source Q_5. The collector current which Q_5 supplies to R_2 equals the biasing current supplied by Q_2 for the configuration shown. Then, once again, the current in R_1 is transferred to the load, except that the current is inverted in this case. Either polarity of input signal results in the same polarity for the load current.

The principal error sources of this circuit are the base current losses of Q_1 and Q_2 and the mismatch between Q_4 and Q_5. From the base current losses a current-sensitive gain error results which equals $1/\beta$. This error is greatly reduced if Darlington pairs are used for Q_1 and Q_2, making the gain error $1/\beta^2$. Additional error is introduced by the base current lost to Q_3, but this error is already limited to $1/\beta^2$. Unequal base currents or emitter-base voltages for Q_4 and Q_5 make the transmission gain of this current source nonunity. For accurate circuit gain, Q_4 and Q_5 must be closely matched for β and emitter-base voltage over the full range of signal currents. Matching over a wide current range is most easily approximated with monolithic dual transistors. Operational amplifier errors add to the limitations of the circuit, but these latter errors are a relatively small addition. Response speed is the primary characteristic limited by the amplifier, as detailed in the next section.

4.3.3 Improving frequency response

High-frequency performance with most of the preceding absolute-value circuits is limited by the speed with which an operational amplifier can turn OFF one rectifying diode and turn ON the other. While the first diode is being turned to OFF, a signal is passed with the wrong polarity, and while the second diode is turning to ON, no signal is passed. For there to be no error during this transition, the transition would have to be instantaneous. However, the slewing rate and gain of the operational amplifier limit the speed with which the output of the amplifier can swing the voltage equal to two diode voltage drops $2V_f$. If the input signal e_i is small, the rate of change of the amplifier output voltage will equal the rate of change of the input signal multiplied by the open-loop gain of the amplifier at the signal frequency $A(f_i)$. The transition time will then be the time required for the input signal to transverse a voltage of $2V_f/A(f_i)$. For larger signals, the rate of change of the amplifier output voltage is limited to its slewing rate limit S_r, and the transition time will be $2V_f/S_r$. Since the ideal transition time would be zero, the response limitations imposed during this time by A and S_r are far more serious than those imposed in simply following the signal. This results in response limits at much lower frequencies for absolute-value conversion than for simple amplifier applications.

Two techniques for reducing these frequency response limitations are presented in this section. The first is a means of removing the response-limiting phase compensation during the switching transition, and the second is a biasing circuit that reduces the transition voltage range. During that portion of the switching transition for which neither rectifying diode is ON, the feedback path around the amplifier is open. In this open-loop state most operational amplifiers require no phase compensation and would have a much higher slewing rate and high-frequency gain without the phase compensation. Then, by switching out the phase compensation for the open-loop state, frequency response limitations are significantly reduced.

For operational amplifiers with external, Miller-effect, or feedback phase compensation,[1] the described operation is achieved by the circuit of Fig. 4.17. Shown here are the switching portions of an absolute-value circuit with conventional feedback phase compensation (a) and the modified compensation (b). With the conventional connection the phase compensation is connected at all times. However, with the modified configuration the amplifier output is connected to a compensation capacitor only when one of the diodes is ON. As a result, phase compensation is connected when it is needed for stability but is not connected when it would unnecessarily limit the switching transition time. This technique can be applied to or adapted to all the previous absolute-value circuits which rely on an operational amplifier to drive the rectifying diodes.

Alternately, the transition time and its associated error can be reduced by decreasing the transition voltage range. As mentioned above, the amplifier output must swing a voltage of $2V_f$ during transition in order to turn one diode OFF and turn the other one ON. This voltage swing requirement can be reduced by biasing the diodes near their ON states. To achieve this without slightly forward-biasing the diodes, the room-temperature bias voltages must be below the ON voltage V_f by $60\,mV$ per decade of current reduction desired in switching to the new OFF state. An $18\,mV$ error in this bias voltage will change the OFF state current by a factor of 2 at room temperature. Not only must the diode bias be accurately placed below a given diode forward voltage V_f, but it must also track the highly temperature-sensitive V_f.

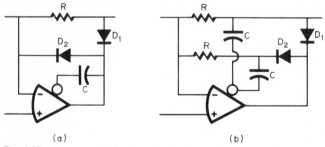

(a)   (b)

Fig. 4.17 Response limitations in absolute-value conversion are reduced by disconnecting the phase compensation during the switching transition.

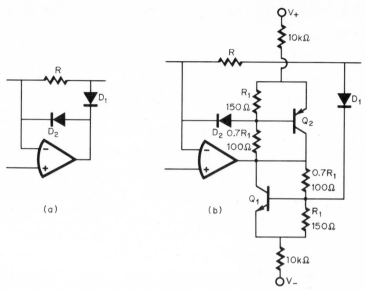

Fig. 4.18 The principal response limitation in absolute-value conversion is reduced by diode biasing for transition swing reduction.

Fortunately, the required bias level and bias tracking can be derived from a matching semiconductor junction, as shown in Fig. 4.18. Here the conventional absolute-value switching section (a) is altered by diode bias networks in (b). In the modified circuit the switching diodes are biased away from the amplifier output level so that both diodes are on the verge of turn-on at the signal zero crossing. The diode bias levels are the voltages on the resistors labeled $0.7 R_1$. No current flows in these resistors from D_1 and D_2 at the signal zero crossing, so that the bias levels are set primarily by currents from the resistors labeled R_1. If the base currents of Q_1 and Q_2 are negligible in comparison, then the currents in the bias resistors are essentially equal to the current V_{BE}/R_1 supplied by the R_1 resistors. This results in diode biases of $0.7 V_{BE}$, and, if V_{BE} approximately matches V_f, the diodes will be biased at 70 percent of their forward voltages. Transition time and its associated errors will be similarly reduced. When switched from the ON state to the biased state, the diode voltages drop by about $0.3 V_f$, or around $180 \, mV$ at room temperature. This decreases the diode currents by a factor of 1,000 to limit OFF state leakage. For the most accurate biasing each diode can be replaced by the emitter-base junction of a transistor which is matched to the associated bias transistor.   The emitter-base junction breakdown voltage does, however, limit signal swing in the latter case.

4.4   Sample-hold and Peak Detector Circuits

In analog instrumentation, sample-hold and peak detector circuits serve as memory elements. With a memory capability computation involving signals which occur at different points in time can be performed, and brief signals can be extended in time. In this way sampled signals are held between updates, and analog signal delays are produced. By holding the output signal of a digital-to-analog converter during switching, a sample-hold circuit removes the transients created by nonsimultaneous bit switching. Most sample-hold circuits can be converted to peak detectors with simple modifications that make the signal determine the operating mode. Specifically, the peak detector circuit switches to a SAMPLE mode whenever the signal exceeds a previously stored level, and then the circuit returns to a HOLD mode whenever the signal drops below the most recently stored level. Generally this operation is achieved by replacing mode-control switches with diodes that are switched by the circuit currents themselves. Some provision is also added for resetting the storage capacitor. Peak detectors are well suited to detection of signal maxima and minima and to amplitude detection with periodic signals.

Discussed in this section are various techniques for improving the accuracy and speed of conventional sample-hold and peak detector circuits.[1] Typically, speed and accuracy are interchangeable in these types of circuits, as they are controlled largely by the size of the storage capacitor. A large storage capacitor reduces the droop with time produced by parasitic current drain on the capacitor, but a small storage capacitor can be charged faster in acquiring a new signal level. The techniques presented here improve performance without making this compromise. Most of the circuit techniques described can be applied to either sample-hold or peak detector circuits.

4.4.1   Accuracy improvements   Typical sample-hold and peak detector circuits are accuracy-limited by the errors of operational amplifiers, of switches, and of the storage capacitor. Finite switching times result in errors considered in the next section. Leakage and dielectric absorption in the storage capacitor are primarily controlled by the choice of capaci-

tor type.   In the SAMPLE mode the stored voltage does not track the signal exactly as a result of signal losses on the control switch and of the normal errors of amplifiers.   In the HOLD mode the stored voltage droops
or decays with time due to parasitic currents, including switch leakage,
amplifier input currents, and capacitor leakage.   Several circuit modifications are described in this section which reduce or compensate for the
errors in the two operating modes.

Some of the simplest sample-hold circuits are variations on resettable
integrators, which were covered in Sec. 3.1.2.   In the RESET mode the
voltage on the integrating capacitor will track the voltage on the RESET
input.   The capacitor will hold its final RESET voltage level when the
RESET circuit is switched out if there is no signal at the integrator input.
Then, by removing the input connection to an integrator and applying the
signal to the RESET input, a sample-hold circuit is formed, as shown in
Fig. 4.19.   With this circuit a mode control is achieved with switch Q_2
which connects resistive feedback around the amplifier in the SAMPLE
mode.

By itself, Q_2 would introduce major errors.   In the SAMPLE mode the
ON resistance r_{ON} of the switch supports part of the voltage intended
for the capacitor.   This error is developed with a current of $I_B +$
$(1/C)(de_i/dt)$, and for slowly varying signals, the error reduces to $I_B r_{ON}$.
In the HOLD mode Q_2 is OFF to disconnect the signal sensing point of
the amplifier.   Now the drain-source leakage current of Q_2 creates error
as it discharges C.   To greatly reduce this discharging the switch Q_1
is added.   This second switch shorts the drain of Q_2 to ground so that the

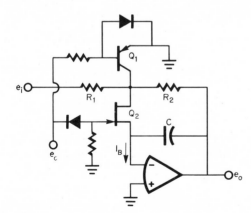

Fig. 4.19 In a simple integrating
sample-hold circuit the leakage
current of mode switch Q_2 is
greatly reduced by holding its OFF
stage voltage near zero, using
switch Q_1.

leakage-inducing voltage across the FET is near zero. In this way the parasitic discharge current is essentially reduced to the small gate-source leakage of Q_1 and the input bias current of the operational amplifier.

For sample-hold circuits the choice between JFET or MOSFET switches is largely dictated by the type of leakage current that can be tolerated. With JFET switches the gate-source leakage current increases drastically at high temperature. While the corresponding leakage current of an MOSFET does not have this high thermal sensitivity, the MOSFET switch does have an equally serious junction leakage current. This is the leakage of the substrate-to-channel junction, and this error current will flow to either the drain or the source or both as controlled by bias voltages. Another significant leakage current not avoided with a MOSFET switch is the drain-source leakage. Both of the leakage currents mentioned can be greatly reduced by maintaining zero voltage drops across their leakage paths.

A means for implementing this leakage reduction technique in a common sample-hold circuit[1] is shown in Fig. 4.20. The basic circuit is modified by the insertion of switch Q_2 and resistor R_2. In the HOLD mode Q_1 is ON for feedback around A_1, and Q_2 and Q_3 are OFF to disconnect the storage capacitor from the input amplifier. Then the voltage on this capacitor is established at the output by voltage follower A_2. A decay in this voltage results from the leakage current through Q_3, which is composed of the gate-drain, drain-source, and drain-substrate leakage currents. Generally, the gate-source leakage is a negligible component, and the other leakages are governed by their supporting voltages.

With the drain-substrate junction across the inputs of A_2, this junction is essentially zero-biased for greatly reduced leakage. Similarly, the drain-source voltage on Q_3 is quite low in this mode, with both Q_2 and Q_3 OFF. The voltage at the source of Q_3 differs from that at its drain by only the drop on R_2 created by the leakage of Q_2 added to the input offset voltage of A_2. Both of the described leakage-reducing biases on Q_3 are made possible by the addition of switch Q_2. This added switch disconnects Q_3 from the input amplifier in the HOLD mode to remove signals that would forward-bias the substrate junction and create drain-source leakage. Without Q_2 the substrate of Q_3 would have to be biased below the most negative signal level, and this bias would result in drain-substrate leakage.

In the SAMPLE mode Q_1 is OFF while Q_2 and Q_3 are ON to connect the input amplifier to the capacitor. Current flow in Q_3 develops a small

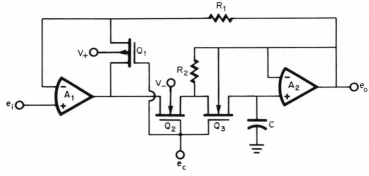

Fig. 4.20 Addition of a third switch to a common sample-hold configuration maintains low voltages on the switch leakage paths that create storage capacitor voltage decay.

voltage with the ON resistance which can tend to forward-bias the source-substrate junction, but any resulting substrate current will not create error under the feedback connection of this mode. With Q_1 OFF the feedback to A_1 is directly from the output. This forces the voltage on the capacitor to remain at whatever level makes $e_o = e_i$ within the input errors of A_1. Essentially no errors are developed by the ON resistances of Q_2 and Q_3 or by the input errors of A_2, as these errors are within the feedback loop. These errors become significant only at high frequencies where the gain of A_1 is low.

The input bias current of an operational amplifier can be as significant an error source in sample-hold circuits as the leakage current of the switch. Further reduction can be made in the HOLD mode voltage droop by means of a circuit that reduces both sources of error current. Once again the associated circuit modifications are illustrated on a common sample-hold configuration (Fig. 4.21). Here the basic circuit is modified by the insertion of switch Q_3 and the feedback capacitor on A_2. As before, Q_1 and Q_2 are switched out of phase to alter the feedback around A_1 and to interrupt the connection to the storage capacitor. However, the HOLD mode decay of the storage capacitor voltage is now compensated by a voltage on the added feedback capacitor. Each capacitor will have a voltage change created by an amplifier input current and by the drain-source and gate-source leakages of the associated switch.

The capacitor voltage changes will be equal and will result in no output voltage droop if the various error currents are matched. Generally, the input bias currents of the operational amplifier will be matched, and

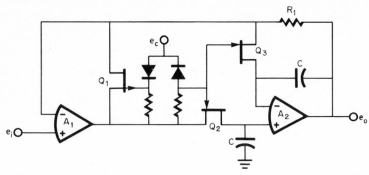

Fig. 4.21 To compensate for the storage capacitor voltage decay during the HOLD mode, a correction voltage is developed on a feedback capacitor.

matched switches will have matching leakages under equal voltages. Since the gates of Q_2 and Q_3 are common, the gate-drain voltages of the switches differ by only the input offset voltage of A_2, and the associated leakage currents will be essentially equal. In this HOLD mode the sources of Q_2 and Q_3 are connected by switch Q_1, so that the drain-source voltages and resulting leakage currents are matched. Straightforward matching of switch leakages and capacitor values will provide an order-of-magnitude reduction in output droop. The compensating capacitor voltage is reset to zero in the SAMPLE mode by the shunting of Q_3. For rapid reset R_1 should be small, but not so small as to heavily load the output of A_2 in the HOLD mode. In HOLD a voltage is developed on R_1 that equals the difference between the input and output signals.

4.4.2 Speed improvements   Ideally, a sample-hold circuit should require only a very short SAMPLE time and should be capable of a very long HOLD time. As described in the previous section, the HOLD time is limited by the output droop resulting from storage capacitor discharge by parasitic currents. The principal determinants of the required SAMPLE time are the transistion times between the two operating modes. For the SAMPLE-to-HOLD transition the aperture time is defined as the time required for switch opening and circuit settling. For the HOLD-to-SAMPLE transition the acquisition time is that necessary for switch closure, capacitor charging, and circuit settling.

Typically, charging the capacitor to the new signal level is a major component of acquisition time, and the charging is rate-limited by either the available current or the slewing rate of the driving amplifier. With a large storage capacitance, the capacitance and the available charging current define the rate limit, yet large capacitances are desirable to reduce droop from parasitic discharge.   Then, the large capacitance needed for long HOLD times increases the required SAMPLE time, resulting in a compromise between the two time characteristics.   One way to avoid this compromise is to use two cascaded sample-hold circuits. In this approach the first circuit has a small storage capacitance for rapid acquisition.   The associated high droop rate is circumvented by the second sample-hold circuit, which acquires the stored voltage from the first circuit before significant droop occurs.   For this transfer the second circuit can have a longer SAMPLE time and a larger capacitance for long HOLD time.

Obviously, another way to increase the capacitor charging rate and decrease acquisition time is to boost the charging current.   To boost the charging current from the driving operational amplifier, the techniques of Sec. 2.2.1 can be applied.   This task is eased by the low duty cycle of the charging current in short SAMPLE times, which permits use of transistors of lower power capability.   In peak detector circuits the current boosting can often be achieved by simply replacing a rectifying diode with an emitter-follower.   Such a modification is shown in Fig. 4.22 for a common peak detector circuit.[1] When the input signal to this circuit slightly exceeds the stored voltage established at the circuit output, the output of A_1 swings positive to turn ON the transistor and further charge C.   The charging current is primarily limited by the collector resistor instead of by the output current capability of A_1.   Charging continues to make the output voltage follow the input signal until this

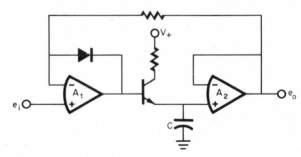

Fig. 4.22   A peak detector with an emitter follower in place of a rectifying diode has a higher limit on the capacitor charging rate.

signal begins to decrease. Since the transistor cannot then supply the opposite polarity current for discharge, it becomes reverse-biased, disconnecting the storage capacitor from the signal. As a result, the capacitor voltage remains at the level of the most recent signal peak unless reset by some discharge mechanism. If the reverse-bias voltage on the transistor is capable of driving the emitter-base junction into breakdown, a protection diode must be added in series with the transistor base.

Another SAMPLE speed limiting factor is the circuit settling time, which affects both acquisition and aperture times. Fast settling is somewhat more difficult to achieve when the circuit feedback loop encloses two operational amplifiers, and will require additional phase compensation. Added phase compensation reduces bandwidth and increases settling time, as discussed in Sec. 1.2.2.

A variety of sample-hold circuits avoid the above settling time increase by using separate feedback loops for individual amplifiers. While this normally leaves the switch ON resistance without feedback for error reduction, this compromise can be avoided with the circuit of Fig. 4.23. In the SAMPLE mode Q_2 connects A_1 to C and Q_3 connects the feedback of A_1 directly to the capacitor. Then, the ON resistance error of Q_2 is within a feedback loop and is greatly reduced. While Q_3 introduces a new ON resistance error, the current in this added resistance is only the input current of the amplifier, so very little error results here also. In the HOLD mode Q_1 connects the feedback of A_1 to the amplifier output instead of to the storage capacitor. Both Q_2 and Q_3 are then OFF to disconnect the storage capacitor from the signal. The added feedback switch Q_3 adds a switch leakage current drain on the storage capacitor

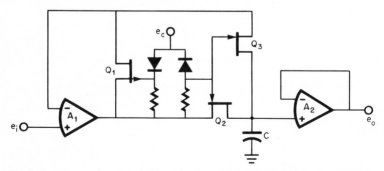

Fig. 4.23 Improved sample-hold settling time is achieved with separate amplifier feedback loops, and the switch ON resistance error can be removed by connecting it in the feedback loop of the input amplifier.

equal to that from Q_2. This independent feedback approach also places the input offset voltage and common-mode error of A_2 outside the SAMPLE mode feedback loop. As a result, these error sources are not avoided as they are in the two-amplifier feedback loop.

4.5   RMS Converters

Several characteristics of a signal can be meaningful measures of its magnitude, including its rms value, its peak value, and its average value. While there are specific cases where the peak or average value is the most direct measure of magnitude, the rms value of a signal is generally the most meaningful since it is an indicator of the energy content of the signal without regard to its waveform. A number of operational amplifier circuit techniques indicate the rms signal level by converting the signal to a corresponding dc voltage. These techniques include analog computing methods and thermal approaches, which use the heating value of a signal as a measure of its energy content.

4.5.1   Computing techniques   By means of analog computation a signal can be electrically processed through the mathematical operations required to derive its rms value. This can be done either by detecting the average value of a given waveform and relating it to the associated rms value or by computing the rms value directly. Although it is limited to a specific waveform, the average reading method is simpler to instrument. To derive the average value of a signal magnitude, it is first rectified by an absolute-value circuit, as discussed in Sec. 4.3, then the rectified signal is filtered to find its average value. The rms signal magnitude is found from this average value using the conversion factor appropriate for the waveform of the signal. To derive this conversion factor the average value of a waveform is compared to the rms value of the waveform, which is defined by

$$E \text{ (rms)} = \sqrt{\frac{1}{T} \int_o^T e(t)^2 \, dt} = \sqrt{\overline{e(t)^2}}$$

For a sinusoid the resulting rms value is 1.11 times the average of the absolute value, and for a bipolar square wave the conversion factor is 1.0. By adding the appropriate conversion factor to the gain of the averaging circuit, an rms-to-dc converter is formed.

In the general instrumentation application it is desirable to detect the

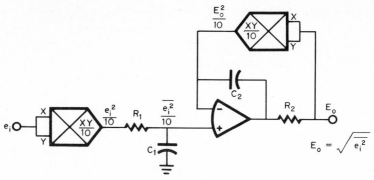

Fig. 4.24 Straightforward analog computation can provide rms-to-dc conversion.

rms value of any arbitrary waveform, and the average reading circuit introduces significant errors when used with other than its design waveform. Even if only one type of waveform is to be measured, distortion components will disturb precise rms measurements. For these reasons it is often desirable to use a true rms reading circuit. To compute the true rms value of a signal the mathematical operations defined in the above expression for rms value can be performed. From this expression it can be seen that the rms computation involves squaring the signal, averaging this result, and then taking the square root of this average.

The circuit of Fig. 4.24 performs these mathematical operations for rms-to-dc conversion. First the input signal is squared by an analog multiplier, then it is averaged by a low-pass filter. To determine the filter element values, the permissible ripple and response time are considered. The output of the filter is applied to an operational amplifier which has a feedback multiplier to form a square-root circuit. To maintain near-zero voltage between its inputs, the amplifier will develop that output voltage which will make the feedback multiplier output voltage equal that of the filter. Since the feedback multiplier is connected in a squaring configuration, the required amplifier output has the indicated square-root relationship with the filter output voltage. The overall result is a dc output voltage equal to the rms value of the input signal. Conversion accuracy is directly limited by the errors of the multipliers and the amplifier, but total low-frequency error can be reduced to 0.1 percent of full scale. Added to the operational amplifier feedback loop are phase compensation elements R_2 and C_2 which may be required to compensate for the phase shift of the feedback multiplier.

4.5.2   Thermal techniques   Since the rms value of a signal magnitude represents its energy content, a direct measure of rms value is the heating value of the signal.   By comparing the heating produced by power dissipation from a signal to that produced by a dc voltage, a null can be attained for which there is equal heating.   At this null the dc voltage will equal the rms value of the signal voltage.   Since the signal can be isolated to a heating element, the rest of the circuit can be connected to process essentially dc voltages, and very wide measurement bandwidths can be achieved.   The key components in this thermal approach are the heating and sensing elements.   For heating elements resistors are convenient because of their ease of fabrication and matching.   Two types of heat-sensing elements are described below; they are thermocouples and semiconductor junctions.

By wrapping a thermocouple in resistance wire, a closely coupled heater/sensor element is formed for rms detection as shown in Fig. 4.25.   As connected, the circuit is a null-seeking feedback loop that develops that output voltage needed to maintain equal thermocouple voltages.   Any signal applied to preamplifier A_1 will increase the dissipation in the input heater resistor.   The resulting temperature change varies the input thermocouple voltage and momentarily unbalances the input voltages of the operational amplifier.   To correct the unbalance, the amplifier develops an output voltage that heats the feedback thermocouple to the same temperature.   This output signal is reduced to a dc level by the filtering of the phase compensation and the slow thermal responses of the heater/sensor elements.   Since the dc output voltage produces the same heating

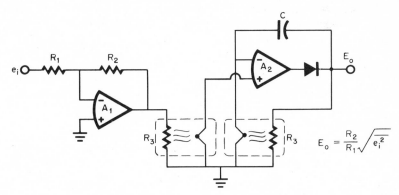

Fig. 4.25   To provide thermal rms conversion, matched heating resistors and thermocouples are connected in a null-seeking comparison circuit.

effect as the input signal, the output voltage equals the rms value of the input voltage. When the input signal is removed, the input thermocouple voltage drops and the polarity of the amplifier output voltage reverses. This reverse-polarity voltage would also create heating, and so it is disconnected by the diode shown.

In addition to the normal accuracy limitations imposed by the amplifiers, the heater/sensor elements introduce several sources of error. Any mismatch in heater resistors results in unequal input and output rms voltages for the equal-heating null. Analogously, any mismatch in the thermal resistances controlling heat losses will result in unequal voltages at the equal-temperature null. A similar error is produced by differences in thermocouple temperature characteristics, which are nonlinear and difficult to match. With care these errors can be reduced to 0.1 percent of full scale. Output settling to this error level is limited to several seconds by the thermal responses of the heater/sensor elements.

Several of the difficulties encountered with the above thermal rms converter can be reduced by using semiconductor junctions instead of thermocouples as the heat-sensing elements.[6] This is accomplished in the circuit of Fig. 4.26 by using the emitter-base junctions of transistors as heat-sensing elements. Due to the gain developed by these sensing transistors, the error effects of the monitoring amplifier are significantly

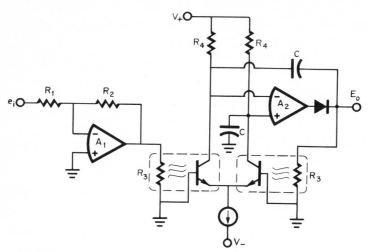

Fig. 4.26 Thermal rms conversion can be achieved with transistors as sensing elements to reduce several errors of the previous thermocouple circuit. *(Koerner, U.S. Patent 3,668,428, 1972)*

reduced over those of the thermocouple form. Circuit operation is analogous to that of the thermocouple rms converter described above. Once again, matched heaters and sensors are connected in a null-seeking feedback loop which reaches equilibrium when the input and output voltages have the same rms value. Under this condition the two voltages create equal heating for equal emitter-base voltages and balanced collector currents.

As with the thermocouple circuit, errors are largely determined by mismatches between the input and feedback signal paths. Unequal input and output rms values result from mismatches in heater resistors, thermal losses, and sensors. In this case thermal losses are easier to control since the resistor and transistor can be integrated together on a small chip. In addition, the low thermal mass of this chip responds faster, improving output settling time. Now the settling time is primarily determined by the filtering phase compensation. Sensor matching is much simpler with semiconductor junctions, as they have linear temperature characteristics which ensure close thermal tracking from matching at only one temperature.[1] Added sources of mismatch error in this circuit are the differences in transistor betas and collector resistors. Overall, the transistor sensor significantly eases reduction of error to 0.1 percent of full scale for this rms converter configuration.

REFERENCES

1. G. E. Tobey, J. G. Graeme, and L. P. Huelsman, *Operational Amplifiers: Design and Applications,* McGraw-Hill Book Company, New York, 1971.
2. J. G. Graeme, Varying Comparator Hysteresis Without Shifting Initial Trip Point, *Electronics,* May 10, 1973.
3. G. A. Korn, Bit Switch for a Calibration-free MDAC, CSRL Memo #201, University of Arizona, Tucson, 1970.
4. J. G. Graeme, Op Amps Form Self-Buffered Rectifier, *Electronics,* Oct. 12, 1970.
5. M. Smither, Improved Absolute Value Circuit, *EEE,* March 1969.
6. H. Handler, A Hybrid Circuit RMS Converter, *Digest ISSCC,* 1971.

5

SIGNAL GENERATORS

Precise, controllable signals of almost any type can be generated by using operational amplifiers.[1] Sine waves, square waves, triangle waves, ramp trains, pulse trains, and timed pulses are generated by circuits of varying complexity and convenience as described in this chapter. In most cases, the signal characteristics are determined by a few components external to the operational amplifiers for ease of control. Variable controls over numerous signal characteristics are also achieved. Voltage control of signal characteristics is outlined for various modulation techniques.

5.1 Sine-wave Generators

As a reference signal, test signal, or carrier, the sinusoidal waveform is the most commonly used. A great variety of circuits have been developed for generating sine waves, including a number of operational amplifier circuits which provide signal precision and circuit simplicity. With operational amplifiers the signal frequency and amplitude can be essentially set by feedback elements alone. Discussed in this section are Wien-bridge oscillators and a variety of other sine-wave generators which are controlled by phase shift, a crystal, a square wave, and multipliers.

5.1.1 Wien-bridge oscillators

For fixed-frequency applications the Wien-bridge configuration is about the simplest precision sine-wave generator. With an operational amplifier as a gain element, such an oscillator will have a frequency that is precisely controlled by the bridge elements. Amplitude control is achieved by means of automatic gain control (AGC), and several such control techniques are described below. Shown in Fig. 5.1 is one of these, where the Wien bridge is formed by the components labeled R and C, and where AGC is provided by zener diode feedback. The Wien-bridge elements supply positive feedback around the amplifier to induce oscillation, and oscillation results at the frequency $f = 1/2\pi RC$, where the positive feedback peaks. A peak in the positive feedback results because the series capacitor increases feedback with frequency while the parallel capacitor decreases it. For equal resistors and capacitors in the Wien bridge, as shown, the peak feedback factor is $\frac{1}{3}$.[1] Then, for a gain of 3 through the amplifier, the gain around the positive feedback loop is unity and oscillation results. The amplifier gain is set by the feedback to its inverting input.

Deviations from unity in the gain around the positive feedback loop cause the oscillation amplitude to diverge or converge with time. Ex-

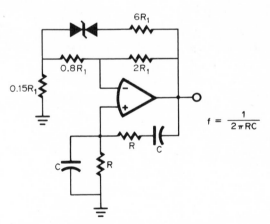

$$f = \frac{1}{2\pi RC}$$

Fig. 5.1 With this Wien-bridge oscillator, the frequency is determined by the positive feedback, and the amplitude is set by the negative feedback.

tremely precise gain setting would be required for amplitude stability without AGC. With AGC the initial gain is set slightly high to initiate oscillation; however, the greater this gain is, the greater the distortion. As the signal amplitude approaches the desired peak level, AGC feedback begins to reduce gain and stops the signal increase. Without AGC the signal amplitude would be limited only by the saturation of the amplifier, and the resulting distortion would be quite high. In Fig. 5.1 the increasing signal turns on the zener diode feedback for the desired gain reduction. A resistor in series with the zener diodes limits the gain reduction to prevent distortion from a drastic gain change. The gain change is further reduced by returning this AGC feedback to the input resistor rather than directly to the amplifier input.

For this sine-wave generator the frequency is determined by the feedback to the noninverting amplifier input and the amplitude is primarily set by the feedback to the inverting input. The accuracy and stability of the waveform frequency are controlled by the characteristics of the Wien-bridge R and C elements. A maximum frequency of operation is set by the slewing rate limit of the operational amplifier. Waveform amplitude will be slightly greater than the zener diode voltage. While the exact amplitude can be varied slightly by adjusting the negative feedback, this also greatly affects distortion. Generally these feedback resistors are chosen for low distortion, and the zener voltage is selected for the desired amplitude. In this way distortion can be reduced to around 0.5 percent with this circuit.

A significant reduction in distortion can be achieved by using an AGC loop that produces the ideal gain level rather than the clamping of the above zener diode approach. The zener clamp technique introduces distortion by the abruptness of the gain change accompanying zener turn-on and by the positive feedback gain of greater than unity needed to initiate oscillation. For more gradual gain change, thermistors and incandescent lamps are often used as signal-sensitive feedback elements. Both types of elements change their resistance to lower gain as their temperatures are raised by increasing feedback signal currents. While this produces a more gradual gain change, the initial gain must still be set high for turn-on.

To also avoid distortion from excess gain, an active AGC loop which reduces gain to the optimum level following turn-on can be used. This is achieved by using an FET as a voltage-controlled resistor, as in Fig. 5.2. The control voltage is derived from the output signal with a common

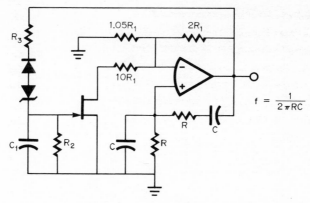

$$f = \frac{1}{2\pi RC}$$

Fig. 5.2 A continuous-acting AGC loop reduces distortion for a Wien-bridge oscillator.

rectifying and filtering detection approach. Here the detected voltage is that which exceeds the turn-on level of the zener diode reference. When the signal peaks are below the detection level, as during turn-on, the FET has zero gate-source bias and appears as a low resistance. This connects the resistor labeled $10R_1$ to temporarily raise gain and initiate oscillation or increase amplitude. As the amplitude passes the detection level, the gate voltage is pulled below that of the source to begin turning the FET OFF and raising its resistance. At some signal amplitude the FET resistance is that which makes the net gain unity around the positive feedback loop, and the amplitude stabilizes. The gain remains near unity throughout the oscillation cycle, as the filter capacitor holds the gate bias between peaks.

In the above operation the critical gain is made greater than unity for turn-on, but the gain is lowered to unity at equilibrium for low distortion. Since the AGC loop controls the gain in both the turn-on and equilibrium states, the adjustment of feedback resistors for gain trim is far less critical than in the previous zener clamp circuit. Some gain change does occur during each oscillation cycle due to the discharge of the filter capacitor C_1 by the reset resistor R_2. By making the associated time constant large, this effect is reduced, but so is the response of the AGC. However, distortion can be limited to 0.2 percent with this active AGC loop.

For a highly precise and stable signal frequency, a parallel mode crystal can be added to the Wien-bridge oscillator of Fig. 5.2, as shown in Fig. 5.3. The very high selectivity of the crystal greatly sharpens the frequency peak of the positive feedback to control the oscillation fre-

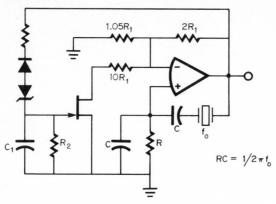

Fig. 5.3   Addition of a crystal to a Wien-bridge oscillator provides highly accurate frequency control.

quency.   In this case the conventional positive feedback elements R and C serve primarily to filter out harmonics of the crystal and must be tuned with the resonant impedance of the crystal.   At resonance the phase shift through the crystal is zero, and so its impedance is resistive.   This resistance replaces one of the resistors of the conventional positive feedback in the circuit shown, although further resistance can be added in series with the crystal.   To match the frequency peak of the conventional feedback with that of the crystal, the resistor R is chosen or trimmed to equal the resonant impedance of the crystal, and the capacitor value C is selected for $RC = 1/2\pi f_o$.   Since the crystal impedance varies with temperature, the gain of the positive feedback loop tends to vary, but the continuous-acting AGC loop will adjust for this.   Where the thermal variation is too great, the gain can be desensitized from the crystal impedance by designing the positive feedback with a resistor in series with the crystal.

5.1.2   Variable frequency circuits   For the Wien-bridge oscillators described above, the frequency-determining elements also affect the critical gain of the positive feedback loop.   Thus, any change in frequency made by altering these components must be followed by a precise adjustment of gain.   For this reason a Wien-bridge oscillator does not permit simple frequency adjustment unless it has a sophisticated AGC circuit to automatically readjust gain over a wide range.   Further discussion of AGC circuits is included in Sec. 6.4.2.   Two other approaches to variable-frequency sine-wave generators which are discussed below are a specialized phase-shift oscillator and a voltage-controlled quadrature oscillator.

The first circuit uses the phase-shift circuit of Sec. 3.5.1 to form a sine-wave generator that has a single resistor frequency control.   To form the oscillator, two of the phase-shift circuits are connected in a loop with an amplitude limiter, as shown in Fig. 5.4.   While the phase-shift circuits, formed with A_1 and A_2, have phase shifts that vary with frequency, the circuits have gain magnitudes which are constant over the bandwidth of the amplifiers.   This independence of gain magnitude and phase responses permits phase response variation for changing oscillation frequency without disturbing the gain magnitude of the feedback loop. The initial gain around the loop is set slightly greater than unity to initiate oscillation.   As the oscillation amplitude rises, the zener diode feedback provides AGC action by lowering gain, as discussed before for Fig. 5.1.

At one frequency the phase shift around the overall feedback loop of Fig. 5.4 reaches $360°$, and this will be the frequency of oscillation. For the phase-shift circuits the phase responses are determined by R_1, R_2, and the capacitors, as expressed by

$$\phi_1 = -2 \tan^{-1}\omega R_1 C - 180° \quad \text{and} \quad \phi_2 = -2 \tan^{-1}\omega R_2 C - 180°$$

Added to these phase shifts is the $180°$ phase shift of the amplitude

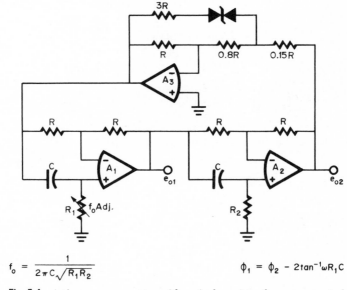

$$f_o = \frac{1}{2\pi C\sqrt{R_1 R_2}} \qquad \phi_1 = \phi_2 - 2\tan^{-1}\omega R_1 C$$

Fig. 5.4 A sine-wave generator with a single resistor frequency control and a second, phase-delayed, output is formed with phase-shift circuits.

limiter, and the combination results in an oscillation at

$$f_o = \frac{1}{2\pi C \sqrt{R_1 R_2}}$$

To vary f_o, either R_1 or R_2 or both can be adjusted. A 10:1 range can be reliably achieved by varying only one of the resistors, as drawn. Beyond this range the phase-shift contribution of one of the phase-shift circuits becomes small, and frequency control is degraded by extraneous small phase shifts. For greater range, R_2 should be switched as a range control while R_1 is used for the variable control. If highly accurate frequency control is needed, R_1 and R_2 should be varied together to keep the phase-shift contributions of both circuits high. A maximum frequency limit is imposed by either the slewing rates or the phase shifts of the operational amplifiers. One unique feature of this circuit is that it will produce two sine-wave outputs with a known, controllable phase difference of

$$\phi_1 - \phi_2 = -2 \tan^{-1} \omega RC \qquad \text{for } R_1 = R_2 = R$$

A signal generator which has electronically variable gain permits automated instrumentation. For this purpose a quadrature oscillator[1] can be frequency-controlled by electronic multipliers,[2] as shown in Fig. 5.5. The principle of operation of a quadrature oscillator is based on the fact that the double integral of a sinusoid yields the negative of a sinusoid of the same frequency and phase. Thus, the result of the double integration can be inverted and used for in phase positive feedback to induce oscillation. With amplitude stabilization, oscillation will occur at that frequency for which the integrator gains are such that integration does not change the signal amplitude. For the elementary quadrature oscillator the oscillation frequency will be that for which

$$\frac{1}{RC} \int \left(\frac{1}{RC} \int A \sin \omega_o t \, dt \right) dt = -A \sin \omega_o t$$

In that case oscillation occurs at $f_o = 1/2\pi RC$.

For the modified quadrature oscillator of Fig. 5.5, the added multipliers vary the frequency of oscillation. Note that one of the multipliers has a negative gain to provide the required inversion in the feedback loop. Alternately, an inverter can be used or feedback can be returned to a noninverting integrator input.[1] With this circuit the output of each integrator is multiplied by $E_C/10$, and this has the same effect as multi-

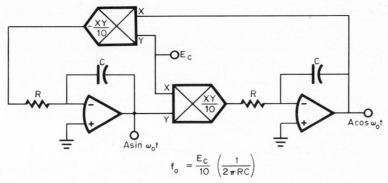

$$f_o = \frac{E_C}{10}\left(\frac{1}{2\pi RC}\right)$$

Fig. 5.5 A sine-wave generator with voltage-controlled frequency is formed by using multipliers to control the gain in the feedback loop of a quadrature oscillator.

plying the integrator gains by the same factor. This alters the frequency at which integration does not change signal amplitude, and thereby alters the frequency at which amplitude-stabilized oscillation will occur. This results in

$$f_o = \frac{E_C}{10}\frac{1}{2\pi RC}$$

As a result, the signal frequency can be varied in direct proportion to the control voltage E_C. To achieve this, some means of amplitude stabilization must be added to the circuit shown. While most of the AGC techniques of Sec. 6.4.2 can be used for this, the computing feedback technique shown later in Fig. 5.7 is attractive for AGC over a wide range of frequencies.

Other advantages are provided by the multiplier frequency control. Very rapid frequency adjustment is permitted by the fast responses of the multipliers. Following a change in E_C, the frequency can stabilize in a fraction of a cycle if a fast-responding AGC circuit is used. With voltage control over frequency, this circuit readily provides frequency modulation. A combined frequency and amplitude modulation is attained if the outputs are taken from the multipliers rather than from the integrators. The range of frequency control is determined by the dynamic range of the multipliers, and a 100:1 range is attainable.

5.1.3 Variable-amplitude circuits For a variable-amplitude sine-wave generator, the previous circuits could be modified with a variable reference for the AGC loop or with a variable-gain output amplifier. In any case, some form of AGC must be used for amplitude stabilization with every sine-wave oscillator. Conventional AGC loops introduce distortion from the inevitable gain variation produced during each cycle. Described below are two other approaches to amplitude control which avoid adding distortion and permit simple amplitude variation. These circuits are a square-to-sine-wave converter and a computing feedback configuration that avoids the frequency sensitivity of common AGC circuits.

With a square-to-sine-wave converter the amplitude control can be applied to the square wave without developing distortion in the sine wave. To make the conversion, the circuit of Fig. 5.6 simply filters out signal components at all frequencies other than the fundamental frequency. Such filtering results in a sine wave at the fundamental frequency with some harmonic components. The harmonics to be filtered out can be determined with a Fourier-series representation of the square wave.

To perform this filtering the circuit of Fig. 5.6 employs a multiple-feedback, bandpass filter[1] preceded by low-pass filter $R_1 C_1$. The bandpass filter is designed for the fundamental frequency by choosing component values to satisfy the equation

$$f_o = \frac{1}{2\pi} \sqrt{\frac{R_2 \parallel R_3}{R_4 C_2 C_3}}$$

Beyond the bandwidth of the operational amplifier, the bandpass filter does not provide the needed filtering. To block the associated high-frequency harmonics, the low-pass filter $R_1 C_1$ is added. When accurately tuned, this circuit will produce a sine wave with less than 1 percent

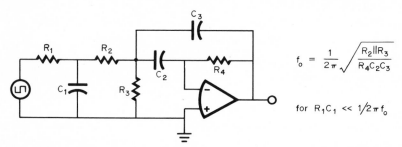

$$f_o = \frac{1}{2\pi} \sqrt{\frac{R_2 \parallel R_3}{R_4 C_2 C_3}}$$

for $R_1 C_1 \ll 1/2\pi f_o$

Fig. 5.6 A variable-amplitude sine-wave generator without distortion from amplitude control can be formed by controlling the amplitude of a square wave and filtering the result.

distortion. Further reduction in distortion can be achieved by adding filter stages identical to the bandpass circuit.

As described earlier for the Wien-bridge oscillators, the distortion introduced by AGC is reduced with a continuous-acting gain control. To make the control continuous, the output signal is rectified and filtered to derive a control voltage. However, the decay of this voltage between charging peaks results in a residual gain change and distortion. In addition, the time constant of the filter is well suited to only a narrow frequency range, so that this AGC technique is not usable in variable-frequency circuits. Both the distortion and the frequency sensitivity of conventional AGC circuits can be removed by using a computing feedback with a quadrature oscillator. The fundamental principle behind this approach is the trigonometric identity

$$\sin^2 \omega t + \cos^2 \omega t = 1$$

Since the quadrature oscillator has both sine and cosine outputs, the computation can be made to provide an amplitude-controlling feedback.

The computing feedback is shown in Fig. 5.7. Here the sine and cosine outputs are squared and compared with a control voltage E_C. At equilibrium the squared signals cancel the control voltage as expressed by

$$A^2 \sin^2 \omega_0 t + A^2 \cos^2 \omega_0 t + E_C = 0$$

Then, the amplitude will be

$$A = \sqrt{-E_C}$$

If the amplitude deviates from this level, the signals do not exactly cancel E_C on the summing network, and a correction signal is developed at the input of comparison amplifier A_3. Then, A_3 changes the voltage it impresses on a gain-control multiplier to correct the amplitude. When equilibrium is restored, A_3 supplies to the gain-control multiplier the voltage that makes its gain -1.

In addition to reducing distortion, this amplitude control improves control accuracy and speed. With a high-gain operational amplifier in the control loop, the amplitude error can be reduced to essentially that caused by the offset and gain errors of the multipliers. In general, the offset of the comparison amplifier and the errors of the comparison resistors can be made negligible. The time required for amplitude change

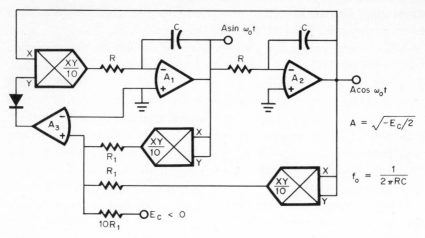

Fig. 5.7 Precise, fast amplitude control is achieved with a computing feedback that makes use of the identity $\sin^2 \omega t + \cos^2 \omega t = 1$.

is now not limited by AGC filters, but merely by the responses of multi-pliers and amplifiers, so that amplitude changes can be made in a frac-tion of a cycle. Also, since the amplitude is controlled by a voltage, this oscillator is suited for automatic instrumentation or amplitude modulation.

5.2 Square- and Triangle-Wave Generators

Square waves and triangle waves are generated simultaneously by opera-tional amplifier circuits through two electronic functions. These func-tions are integration and voltage comparison with hysteresis. Integration of the output of a comparator in one of its states develops a ramp, and this ramp can be used to trigger the comparator to reverse its state. In the reversed state the signal integrated will be of opposite polarity, and so the resulting ramp will be of opposite slope. With hysteresis in the comparator, the new ramp will cause reversal at a different trigger point, and at this point the original state is resumed. As this action continues, the comparator output traces out a square wave, and the in-tegrator output is a triangle wave. Described in this section are several circuits which generate these signals with only one operational ampli-fier and additional circuits which use two amplifiers to improve precision and provide variable waveform controls.

5.2.1 Single amplifier configurations

Both the integration and voltage comparison required for generating square and triangle waves can be performed by one operational amplifier. This is achieved by using a simple resistor/capacitor integrator as a feedback network to the input of a comparator circuit, as in Fig. 5.8. Here R and C perform the integration and the amplifier compares the voltage on C with the reference established at its other input by the hysteresis feedback of R_1 and R_2. When the capacitor voltage integrates up to the level of this feedback reference, the comparator state reverses to reverse the slope of the integration voltage and to reverse the polarity of the feedback reference voltage.

The continuation of this action produces a moderately precise square wave at the comparator output. To fix the amplitude of this signal the dual zener diode clamps the output, with current-limiting provided by R_3. The rise time is set by the slewing rate of the amplifier. Determining the frequency of the signal are the integrator time constant and the hysteresis feedback as expressed by

$$f = \frac{1}{2RC \ln(1 + 2R_1/R_2)}$$

This frequency can be varied by means of resistor R without altering the amplitude of either signal output. Note that the frequency as expressed is independent of the square-wave amplitude even though the

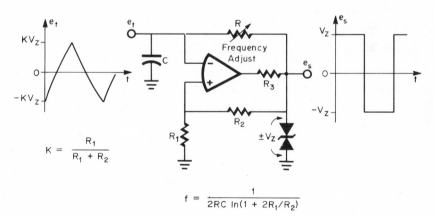

$$f = \frac{1}{2RC \ln(1 + 2R_1/R_2)}$$

Fig. 5.8 Integrating feedback around a comparator circuit produces an oscillation with square- and triangle-wave outputs.

amplitude determines the capacitor charging rate. This is because the same amplitude determines the level of the hysteresis feedback. If a variable output clamp were used, the circuit would have amplitude control that would be independent of the frequency control.

For the simple integrator used in this circuit the slope of the ramp is not linear, but rather it is a portion of an exponential capacitor charging characteristic. Improved linearity is achieved by using only the initial portion of this charging characteristic, as occurs when the comparator trip points are well below the output voltage levels. To establish this condition, the comparator hysteresis should be made a small portion of the output voltage, as described by the fraction

$$K = \frac{R_1}{R_1 + R_2}$$

Note that the linearity of the triangle wave, as well as the oscillation frequency, can be seriously affected by any input current drawn by the operational amplifier. Such current will flow into many operational amplifiers under the differential input overload created by the operation of this circuit. With amplifiers which have input protection circuits that lower input resistance, large resistors should be added in series with each amplifier input.

Far better triangle-wave linearity can be achieved by maintaining a constant capacitor charging current magnitude, as this results in a constant rate of change of voltage with time. In the above circuit the charging current decreases as the capacitor voltage increases, but this can be avoided by replacing the charging resistor with current sources, as in Fig. 5.9. In this circuit the capacitor charging feedback consists of FET current sources that will supply both polarities of charging current. For a given polarity of output voltage, one gate will be reverse-biased for current source operation, and the other gate will be forward-biased, operating simply as a diode in series with the drain of the current source. When the output polarity switches, the roles of the two FETs are interchanged, and the polarity of the charging current reverses. Now the frequency of oscillation is controlled by the charging current i_c, as related by

$$f = \frac{i_c}{4KV_ZC}$$

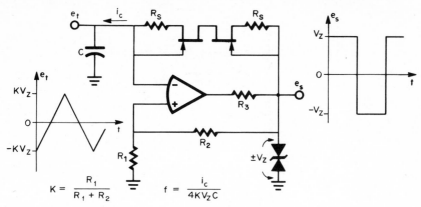

Fig. 5.9 Improved triangle-wave linearity is provided by current source charging of the capacitor.

As indicated by the term V_Z above, this circuit modification makes the frequency dependent on amplitude. Otherwise, this circuit is identical to that of Fig. 5.8.

Since the FETs will operate as floating current sources, their bias is fairly simple. No separate reference voltage is needed to set their current levels. However, for thermal stability, the bias-setting resistors labeled R_S should be chosen to set the current levels near the zero temperature coefficient level of the FETs. In addition, the FETs should be matched to ensure waveform symmetry. To operate as current sources, the FETs must have a gate-drain voltage large enough to induce pinchoff. Then, the triangle-wave amplitude must be less than that of the square wave by a voltage that at least equals the sum of the pinch-off voltage and the forward gate drop. It is the residual current dependence on gate-drain voltage that is now the major source of nonlinearity in the triangle wave. This dependence is expressed by the current source output resistance of $R_O = (1 + g_{fs}R_S)r_{ds}$, and the associated nonlinearity is typically around 1 percent. This nonlinearity can be further reduced by increasing R_O with cascode bias on the current sources. The significance of the operational amplifier input current is greater in these attempts to improve linearity. For this reason the amplifier used should retain high input resistance under differential input overloads. Note that any load current drawn from the triangle-wave output also disturbs linearity, and buffering may be necessary.

Another variation on the signal generator considered here is one that adds a synchronizing input for timing coincidence. To achieve synchro-

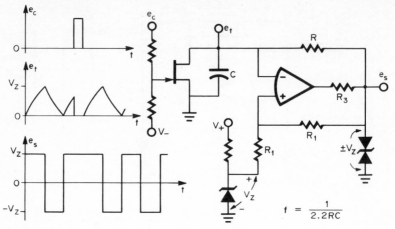

Fig. 5.10 To synchronize the square- and triangle-wave generator with another signal, the hysteresis feedback is shifted to make zero an initial capacitor voltage, and the timing capacitor is reset to zero voltage.

nization, the timing capacitor voltage must be reset to the initial value it normally has at the beginning of one half cycle. With the circuit of Fig. 5.10, the capacitor voltage is reset to zero when a control signal e_c turns ON the FET switch. This resets the circuit to the beginning of the positive half cycle of the square wave coinciding with the end of the control pulse. To make the zero volt RESET level one of the capacitor initial voltages, a constant voltage is added to the hysteresis feedback. The added voltage is supplied by a zener diode as shown, and it provides the desired voltage shift as long as its voltage matches that of the dual zener. Any mismatch in these voltages will result in a difference between the capacitor RESET voltage and its true initial voltage. The effect of this RESET error can be trimmed out by unbalancing the hysteresis feedback resistors. In fact, by varying this feedback, the signal generated can be made to either lag or effectively lead the control signal. Another source of RESET error is the residual capacitor voltage left because of finite discharge time. The discharge is controlled by a time constant of $r_{ON}C$, where r_{ON} is the ON resistance of the FET.

5.2.2 Variable circuits By using separate operational amplifiers for the integration and the voltage comparison, waveform control and greater precision are added to square- and triangle-wave generators. Described below are generators which have potentiometer controls for virtually every waveform characteristic and which have a multiplier for electronic control of frequency. A more precise triangle wave and more accurate frequency control are attained by using a separate amplifier for the integrator. The first circuit is basically an integrator with a feed-back comparator that switches the reference voltage to be integrated,[3] as shown in Fig. 5.11. Forming the integrator are A_1, R_f, and C. The voltage integrated is the difference between those supplied by the com-parator output and the symmetry control, or $e_s - V_S$. Since e_s switches polarities, V_S alternately increases and decreases the voltage integrated. This increases one integration rate and decreases the other to control the waveform symmetry, as expressed by the ratio of the two integration times in the form

$$S = \frac{V_Z - V_S}{V_Z + V_S}$$

Symmetry is then affected only by control potentiometer R_S. The ac-curacy of this control is primarily set by the accuracies of R_S and its biases, while the errors produced by the input bias current and offset

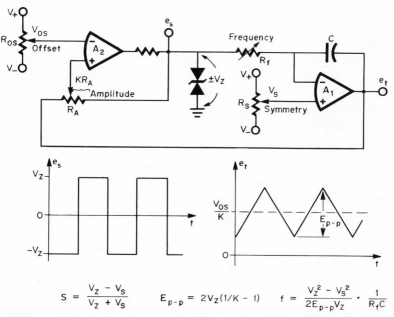

$$S = \frac{V_Z - V_S}{V_Z + V_S} \qquad E_{p-p} = 2V_Z(1/K - 1) \qquad f = \frac{V_Z^2 - V_S^2}{2E_{p-p}V_Z} \cdot \frac{1}{R_f C}$$

Fig. 5.11 Comparator feedback around an integrator produces precise square and triangle waveforms that are highly controllable.

voltage of A_1 are negligible in most cases. Control range is limited by the maximum integration rate, set by either the operational amplifier slewing rate or the charging rate limit of C, set by the output current of A_1.

As the output of A_1 traces out the triangle wave, the waveform peaks are set by the trip points of the comparator. The comparator is composed of amplifier A_2, the zener diode output limiter, and hysteresis feedback through R_A. Switching occurs when e_t overcomes the hysteresis feedback from e_s to raise the noninverting input of A_2 to the offset voltage set by R_{OS}. Both the positive and the negative switching points are affected by the offset control, so that it merely shifts the midpoint of the waveform, as expressed by $V_{mid} = V_{OS}/K$. The controls which determine the midpoint are R_{OS} and R_A, which sets K as illustrated. The offset accuracy is essentially determined by the precisions of R_{OS}, R_A, V_+, and V_-, with generally negligible errors contributed by the input offset voltage and bias currents of A_2.

In determining the comparator trip points, the hysteresis feedback signal of R_A is combined with V_{OS}. Hysteresis is established by the alternating polarity of the voltage supplied through R_A, and this separates the two trip points to establish the triangle-wave amplitude at

$$E_{p-p} = 2V_Z (1/K - 1)$$

The amplitude is then controlled only by the setting of amplitude control R_A, as expressed by K. Both the accuracy and the stability of the amplitude are set by those of R_A and V_Z. The range of amplitudes attainable is limited to below the output swing limits of A_1 and above the comparator input offset voltage.

Since the amplitude determines the voltage change of the integration, it directly affects the time of integration and thus the waveform frequency. The other factors determining frequency are the two integration rates, as set by the time constant R_fC, and the input voltage $e_s - V_S$ of the integrator. Considering these factors, the frequency is found to be

$$f = \frac{V_Z{}^2 - V_S{}^2}{2E_{p-p}V_Z} \; \frac{1}{R_fC}$$

While the frequency is directly varied by the control R_f, the amplitude and symmetry controls also create variations. The accuracy and stability of the frequency are set by the accuracy and stability of the elements of the above expression. The upper limit is set by the maximum integration rate, controlled by the slewing rate limit of A_1 or by the charging rate limit of C from A_1. A lower limit is imposed at a very low frequency by the integration error of the input bias current of A_1.[1]

Once again, a multiplier can be used to provide electronic control for automatic signal variation. For electronic frequency control of a square- and triangle-wave generator, a multiplier is inserted in the generator feedback loop, as shown in Fig. 5.12. Otherwise, this circuit consists of an integrator with a feedback comparator as described above. With the multiplier the amplitude of the signal to be integrated is multiplied by control voltage E_C, and this multiplies the rate of integration or the triangle-wave slope by the same factor. As a result, the integration time between comparator trip points is reduced, and the oscillation frequency becomes related to E_C by

$$f = \frac{E_C}{20R_C}$$

The accuracy of this control is primarily limited by the nonlinearity and the offset of the multiplier. Very rapid frequency changes can be made,

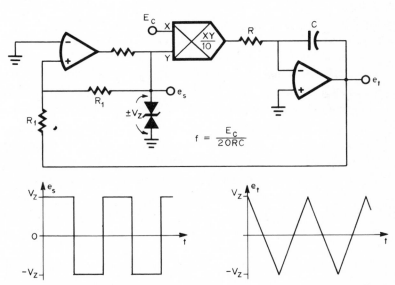

Fig. 5.12 Electronic frequency control is achieved with a multiplier in the feedback loop of a square- and triangle-wave generator.

as permitted by the responses of the elements in the feedback loop. Any mismatch in the voltages of the dual zener diode will disturb the symmetry of the waveforms, but this can be corrected by a trim resistor to the comparator feedback point from a dc bias or by a diode bridge limiter.[1]

Electronic amplitude control can also be produced with multipliers. The multiplier shown varies the square-wave amplitude, but it cannot do so without affecting frequency. Analogously, a multiplier between the integrator output and the comparator input varies the triangle-wave amplitude along with the frequency. For independent amplitude control, multipliers can be used to simply multiply the output waveform amplitudes.

One further waveform characteristic that can be readily controlled with an operational amplifier square- and triangle-wave generator is the shape of the triangle wave. In particular, the normally linear slopes of the triangle wave can be made exponential, with either a positive or negative exponent. With this control the nonlinearities of loads driven by a triangle wave can sometimes be compensated by the appropriate exponential wave shape. Such a requirement can arise when driving a long line that has a capacitive load.

While a negative exponent can be developed with a simple resistor capacitor lag network, a positive exponent requires positive feedback. By increasing the positive feedback supplied to a lag network, the exponent can be varied from negative values through zero to positive values. This exponent control is achieved with the square- and triangle-wave generator of Fig. 5.13. Once again, integration and voltage comparison combine to generate the waveforms. The comparator shown is the same as that of previous circuits, and it again produces a square wave that swings from V_Z to $-V_Z$. However, the integration function is provided by a circuit with the desired exponential response control.[4] Without R_5 this circuit would have the decreasing exponential response of the lag network formed by R_4 and C. With R_5 a positive feedback is supplied to C, and it tends to make the triangle-wave shape that of an increasing exponential.

By combining the responses of the lag network and the positive feedback, an exponential response with an exponent of any magnitude and of either polarity can be produced. The shape of the triangle wave generated is described by

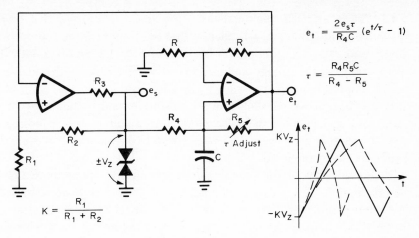

Fig. 5.13 A triangle wave with a variable exponential shape is generated by using positive feedback to a lag network.

$$e_t = \frac{2e_s\tau}{R_4C}\,(e^{t/\tau} - 1) \qquad \text{where } \tau = \frac{R_4R_5C}{R_4 - R_5}$$

From the expression for τ it is seen that the exponent becomes positive when R_5 becomes less than R_4. When the two resistors are equal, the triangle wave is linear with a frequency of

$$f = \frac{1}{4KR_4C} \qquad \text{where } K = \frac{R_1}{R_1 + R_2}$$

The signal frequency varies with the response exponent, since the capacitor charging speed is changed.

5.3 Ramp and Pulse Generators

A ramp waveform is useful for sweep testing and display, and generation of this waveform often also produces a pulse train that aids in timing and sampling. To build a ramp and pulse generator, one half cycle of a square- and triangle-wave generator can be shortened. In this way the circuits of Sec. 5.2 can be converted for ramp and pulse outputs. Generally, the conversion is accomplished by greatly reducing the time required to charge the timing capacitor in one direction. Described in this section are several such converted circuits and a staircase ramp generator.

5.3.1 Single amplifier configurations
To generate a ramp waveform a voltage is integrated on a capacitor, and this ramp can be sensed by a comparator which produces a pulse in resetting the circuit. Both the integration and the voltage comparison can be performed with one operational amplifier in several ways to be discussed. The first approach, shown in Fig. 5.14,[5] parallels that of the square- and triangle-wave generator of Fig. 5.8. With the pulse width control potentiometer indicated, a wide range of duty cycles can be accurately developed; however, the ramp output is nonlinear.

Essentially, the circuit is an astable multivibrator having its two characteristic time intervals determined by separate feedback elements. Astable operation is ensured by positive feedback through R_3 and R_4. From this feedback the switching threshold voltage is set at the non-inverting input. Switching occurs when the inverting input voltage is brought to this threshold level by the charging of capacitor C. The capacitor charging current is supplied from e_p by one of the feedback resistors, R_1 or R_2. A positive e_p forward-biases D_1, and R_1 controls the rate of charging with time constant R_1C. When e_r reaches KV_Z, the output switches, reversing the polarities of the charging current and the positive feedback. The negative output then discharges C through R_2 with a time constant that is now R_2C. Since separate resistors control

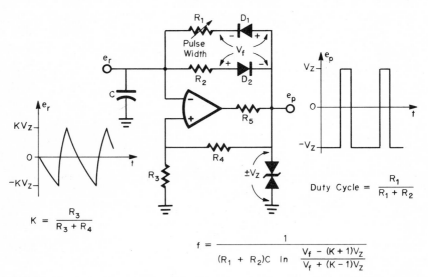

$$K = \frac{R_3}{R_3 + R_4}$$

$$\text{Duty Cycle} = \frac{R_1}{R_1 + R_2}$$

$$f = \frac{1}{(R_1 + R_2)C \ \ln \dfrac{V_f - (K+1)V_Z}{V_f + (K-1)V_Z}}$$

Fig. 5.14 By means of separate feedback resistors for the two output states, this ramp and pulse generator provides a controllable duty cycle.

the time intervals of the two output states, the duty cycle is set by simply choosing the resistors, and it will be $R_1/(R_1 + R_2)$.

In practice the range of duty cycles attainable is limited by the slewing rate, output current, and input bias current of the amplifier. Slewing rate controls the large-signal rise time and, thereby, limits the minimum time interval of either state. This minimum is also limited by the slewing rate of the capacitor as set by the amplifier output current available for charging. With the minimum interval determined by the preceding two factors, the maximum interval length is limited by input bias current. This input current must be small compared to the charging currents supplied by R_1 and R_2. For this reason high input resistance must be maintained by the operational amplifier under input overload. The ramp output is nonlinear since it is generated by means of capacitor charging through a resistor.

To improve the linearity of the ramp waveform, the appropriate capacitor charging current can be made constant with a feedback current source. This is achieved with the same technique that was used to linearize the triangle wave of Fig. 5.9. The result is the circuit of Fig. 5.15, and it has the same basic astable operation as described above except for the capacitor charging. Now the capacitor is charged by an FET which supplies a constant current for charging and a low impedance path for discharging. When the amplifier output voltage is positive, the gate junction of the FET is reverse-biased for operation as a constant current source. However, when the amplifier output voltage swings negative, the gate junction forward-biases, providing a diode feedback for rapid discharge. The speed of the discharge is determined by the current available from the amplifier output through resistor R_3. Assuming that the discharge time is small, the generated signals have a frequency of

$$f = \frac{i_s}{2KV_ZC}$$

To ensure a linear, stable ramp the normal limitations of an FET current source must be observed. Since an FET current source will float without a separate reference voltage, it is very convenient for this application. However, a temperature-stable current output requires that biasing resistor R_S be selected to set the FET at its zero temperature coefficient point. In addition, a minimum bias voltage must be maintained across the FET to keep it in pinchoff and maintain a constant output

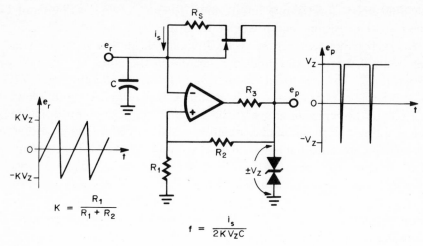

Fig. 5.15   A linear ramp output is produced by constant capacitor charging current from an FET current source.

current. Even when this voltage bias condition is maintained, the charging current supplied will vary slightly with the change in voltage across the FET. This effect is represented by the current source output resistance, which is

$$R_O = (1 + g_{fs}R_S)r_{ds}$$

Further degradation of ramp linearity is produced by amplifier input current and load current from the ramp output. To avoid this degradation the amplifier must maintain high input resistance under input overload, and output buffering may be needed.

Yet another ramp and pulse generator circuit can be formed with a single operational amplifier, and this one avoids the need for a separate output buffer. In this case the circuit is an integrator with a voltage-sensing RESET switch, as represented in Fig. 5.16.[6] By integrating a constant voltage, a constant capacitor charging current produces a linear output ramp. The output voltage rise continues until Q_1 and Q_2 turn ON, as initiated by the emitter-base breakdown of Q_2. Breakdown results from the inverted connection of Q_2 and provides a low output voltage limit expressed by

$$e_{r\,max} \overset{\Delta}{=} V_P = -V_f + V_{CB} - BV_{EB} + V_{BE}$$

Since the collector-base junction of inverted Q_2 is forward-biased, its voltage drop is that of a forward-biased junction and $-V_f + V_{CB} \doteq 0$. Then

$$V_P \doteq -BV_{EB} + V_{BE} \approx 7 \text{ V typically}$$

This peak voltage is quite temperature-stable, as the typical $+2.6$ mV/°C sensitivity of $-BV_{EB}$ is largely canceled by the -2mV/°C variation in V_{BE}. As a result, the temperature coefficient of the output peak V_p is around $0.01\%/°C$. Since the signal is a linear ramp, frequency drift is also $0.01\%/°C$. Frequency is set by the integration time at

$$f = \frac{V_-}{V_P} \frac{1}{RC}$$

Once the emitter-base junction of Q_2 breaks down, a base current is supplied to Q_1. Then Q_1 in turn supplies base current to Q_2 and discharging current to the capacitor. As the capacitor voltage drops below the breakdown point, Q_1 and Q_2 remain ON from their positive feedback to continue discharging C. Discharging continues until the capacitor voltage will no longer sustain V_{BE}. At this point the capacitor voltage equals V_{BE} and the output is

$$e_{r\,min} = -V_f + V_{BE} \doteq 0$$

The canceling voltage provided by V_f thereby removes the output offset.

With the above circuit, the ramp-train frequency is limited only by the slewing rate of the operational amplifier. Ramp linearity is limited by

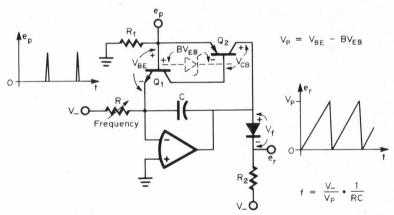

Fig. 5.16 A buffered, linear ramp output is provided by an integrator with a voltage-sensitive RESET switch.

leakage from Q_1 and Q_2, the sensitivity of V_f to e_r, and an amplifier input overload recovery following discharge. Leakage from Q_1 and Q_2 is limited by R_1, which also prevents this leakage from turning the transistors ON. A stray turn-on current is also created by capacitance coupling from the output, making the required value of R_1 frequency-dependent. Linearity error from variations in V_f is a result of the change in its current as e_r changes the drop on diode bias resistor R_2. Nonlinearity from these effects is about 0.3 percent.

5.3.2 Other ramp and pulse generators

By using two operational amplifiers to form a ramp and pulse generator, greater waveform control and precision is achieved. Described below are a circuit having potentiometer control over most waveform characteristics and a circuit that generates a staircase ramp. The first circuit is quite similar to the triangle- and square-wave generator of Fig. 5.11. Once again, an integrator and a voltage comparator are connected in a feedback loop, but in this case they are connected so that the comparator forces a rapid negative integration, as shown in Fig. 5.17. Only the negative integration is influenced by the comparator, as the diode shown disconnects the compara-

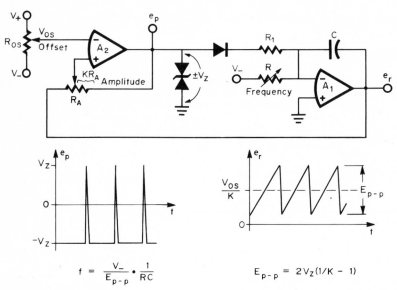

$$f = \frac{V_-}{E_{p-p}} \cdot \frac{1}{RC}$$

$$E_{p-p} = 2V_Z(1/K - 1)$$

Fig. 5.17 An integrator and a comparator generate a controllable, highly linear ramp and an accurately timed pulse train.

tor during the positive integration. By making R_1 much smaller than R, the fall time of the ramp will be made much less than the rise time.

With the circuit of Fig. 5.17, the waveform is largely controlled by three potentiometers. To set the amplitude of the ramp, the comparator trip points are adjusted by means of R_A to give

$$E_{p-p} = 2V_Z\left(\frac{1}{K} - 1\right)$$

By varying the midpoint of the comparator trip points with R_{OS}, the waveform is offset by a voltage of V_{OS}/K. The ramp frequency is varied by adjusting the integration resistor as related by

$$f = \frac{V_-}{E_{p-p}} \frac{1}{RC}$$

A minimum fall time is imposed by the limited currents available from the amplifier outputs to discharge the capacitor. To avoid this limit the comparator can instead be used to drive an FET RESET switch connected across C.

A staircase waveform is a specialized ramp that is useful for sequential control and multiple-level testing. Such a signal can be precisely produced with a D/A converter driven by a clock-controlled counter. A much simpler staircase generator which retains moderate precision can be formed with operational amplifiers, as shown in Fig. 5.18.[7] As will be described, this circuit forms a staircase by differentiating and reintegrating a square-wave train with a rectification that eliminates one transition of the square wave.

Amplifier A_1 performs the differentiation and rectification. Square wave e_i, which can be produced with the circuits of Sec. 5.2, is impressed on C_1 to develop a current $i = C_1\,de_i/dt$ for R_1 small. For the positive transistion this current is positive, and D_1 conducts with Q_1 and Q_2 OFF. For the negative transition the conductions reverse, and Q_1 and Q_2 conduct the derivative current. From this rectification action only the negative-going transition transfers a current through Q_1 and Q_2 to the second amplifier. Voltage bias across these transistors is provided by biasing one input of A_1 below ground through R_2 and R_3.

Integration of the current supplied by Q_1 and Q_2 is performed by A_2 to produce the staircase output. This current flows in integrating capaci-

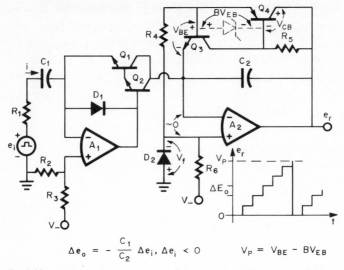

$$\Delta e_o = -\frac{C_1}{C_2}\,\Delta e_i, \ \Delta e_i < 0 \qquad V_P = V_{BE} - BV_{EB}$$

Fig. 5.18   To form a staircase, A_1 differentiates a square wave and passes the derivative current of the negative transition to A_2, where it is reintegrated.

tor C_2 to produce an output voltage change of

$$\Delta e_o \doteq \frac{-1}{C_2}\int i\,dt \qquad \text{for } i < 0 \text{ and } \beta_1\beta_2 \gg 1$$

Replacing i with its expression above gives

$$\Delta e_o \doteq -\frac{C_1}{C_2}\,e_i \qquad \text{for negative transitions}$$

Or more simply, Δe_o is related to the input transition Δe_i by

$$\Delta e_o \doteq -\frac{C_1}{C_2}\,\Delta e_i \qquad \text{for } \Delta e_i < 0$$

Each negative transition of the input square-wave train creates a step in the output voltage, producing a staircase waveform.

Output steps continue until the output voltage reaches the trigger level of the RESET clamp. This clamp is described for the circuit in Fig. 5.16, and it triggers when the emitter-base junction of Q_4 breaks down, limiting the output to a peak of

$$V_P \doteq V_{BE} - BV_{EB} \qquad \text{as } V_{CB} \doteq V_f$$

This peak is quite temperature-stable, as the thermal variations of V_{BE} and BV_{EB} largely cancel, and the resulting temperature coefficient of V_p is around $0.01\%/°C$. The breakdown supplies base current to Q_3 and turns ON the positive feedback loop it forms with inverted transistor Q_4. With this conduction C_2 is discharged until its voltage reaches V_{BE}, below which the clamp conduction is not sustained. At this point, the output voltage has been returned to $e_o \doteq V_{BE} - V_f \doteq 0$. The clamp then turns OFF and a new staircase begins.

Gain errors and nonlinearity vary with the operating frequency and component characteristics. The input currents of the amplifiers remove some of the derivative current pulses, and they also create output sag by discharging C_2. To avoid severe error from these input currents, amplifier input protection which draws high current during input overload must be avoided. Overload recovery and slewing rate limit the operating frequency range. Further current transfer error results from the finite betas of Q_1 and Q_2. However, for this Darlington pair the loss is only about 0.002 percent. Leakage of the RESET clamp introduces an output nonlinearity which is reduced by R_4 and R_5. The endpoints of the staircase are controlled by the voltage relationships described earlier. With these error sources the gain error is typically 0.05 percent, and the nonlinearity will commonly be 0.1 percent.

Several other ramp and pulse generators can be formed by circuits similar to the triangle- and square-wave generators of Sec. 5.2. The electronic frequency control achieved with a multiplier in Fig. 5.12 can be applied in an analogous manner to the ramp and pulse generator of Fig. 5.17. Once again, the multiplier is connected in series with the frequency-control resistor, and it sets the voltage to be integrated. A ramp with a controllable exponential shape can be generated with the circuit of Fig. 5.13 if a diode is connected in parallel with the control potentiometer.

5.4 Monostable Multivibrators

A monostable multivibrator is a widely used timing signal generator that can be formed with an operational amplifier for timing precision and stability. As will be described, these monostable circuits can be designed for either reset capability or reset immunity during the timing pulse and for sequential generation of two independent pulses. In the simplest case, the monostable multivibrator is formed by a voltage com-

parator that compares the voltage on a charging capacitor with a reference voltage, as in Fig. 5.19. Here the comparator is an operational
amplifier, and the reference voltage V_t sets the switching threshold.
To initiate the timed output pulse, a trigger signal e_t turns ON the FET
switch and resets the capacitor voltage to zero. This places the inverting
amplifier input at a voltage below the noninverting input so that the output swings positive. Then, the capacitor charges through resistor R
to raise the voltage on the inverting input and reverse the output state
when this voltage reaches the threshold voltage V_t. The charging time
from the end of the trigger pulse until V_t is reached will essentially be
the pulse length if trigger pulses are short. Under this condition the duration of the pulse will be

$$t_p = RC \ln (1 + R_2/R_1)$$

By means of potentiometer R a linear control over t_p is achieved.

Timing accuracy is controlled by only a few components with this
circuit. Only minor errors will be introduced by the gain error, input
offset voltage, and input bias currents of the operational amplifier, as
long as it maintains high input resistance under overload. If overloads
lower the input resistance, as occurs with some input protection techniques, large resistors should be connected in series with the amplifier
inputs. Otherwise, the primary sources of timing error are the inaccuracies of the passive component values. Some additional error will result if the trigger pulse is either too short to allow capacitor voltage
RESET or if it is significant in length compared to t_p. However, this
circuit does avoid the customary error associated with triggering most

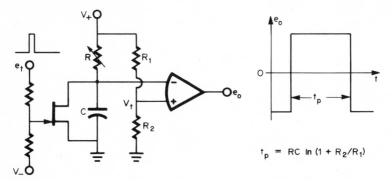

Fig. 5.19 A precise, variable monostable multivibrator is formed with an
operational amplifier that compares the voltage on a charging capacitor with a
reference.

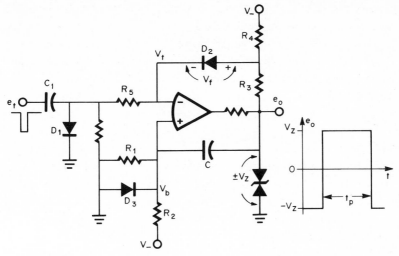

Fig. 5.20 By means of a feedback clamp this monostable multivibrator is made immune to trigger signals that continue into or occur during the timed interval.

monostable multivibrators before a recovery time has elapsed. While most such circuits require time to recover between timing pulses, this circuit can be retriggered any time after or even during a timing pulse. In every case the trigger causes the timing to start at zero.

In some applications it is undesirable to have the timed pulse affected by trigger signals that continue into or occur during the timed interval. This is true where the time added by the trigger pulse duration would be significant in comparison with the duration of the output pulse. For these applications the trigger input can be disabled during the timed interval by using the monostable multivibrator of Fig. 5.20.[8] Once the circuit has been triggered, the resulting positive output voltage clamps the trigger input through D_2. In this way the inverting amplifier input is held at a voltage that sets the switching threshold, and the voltage is

$$V_t \doteq \frac{(V_Z - V_-)R_3}{R_3 + R_4} - V_f \qquad \text{for } R_5 \gg R_3 \parallel R_4$$

No signal at the trigger input will alter this threshold to affect timing as long as the clamp holds. Since the clamping might be upset by fast positive trigger pulses, such pulses are suppressed by diode D_1.

Monostable action results from feedback through the timing capacitor C. When the circuit is triggered by a negative-going pulse, the output swings positive and supplies positive feedback through C to hold the output in this state. But C discharges through R_1 and R_2 until it lowers the noninverting input voltage to the threshold level V_t, and then the output switches back to the negative state. Now the capacitor charging is reversed to return the noninverting input to its equilibrium voltage. This voltage is set by the divider action of R_1 and R_2 at

$$V_{bo} = \frac{R_1 V_-}{R_1 + R_2}$$

Until this equilibrium voltage is reached, the circuit cannot be retriggered without encountering timing error, and so this circuit is limited by a recovery time. This recovery time is greatly reduced by adding D_3 to provide a low impedance charging path in this mode. To ensure that this diode does not alter V_{bo}, the voltage is set well below the forward-bias level of the diode. In practice, V_{bo} need only be large enough to hold the output negative in the presence of noise and offset. By the above action, the timing pulse has a duration of

$$t_p = (R_1 \parallel R_2)C \ln \frac{2V_Z}{V_t + V_{bo}}$$

Another monostable multivibrator configuration produces two sequential timing pulses.[9] The two pulses are of opposite polarity and have independent time durations, so that they can be used to time two separate, sequential operations. Once again, monostable timing action is achieved by positive feedback through the timing capacitance, as described for the last circuit. In the circuit of Fig. 5.21, two such feedback paths are provided. Each feedback path controls timing for only one polarity of output voltage, as controlled by D_1 and D_2. By means of negative feedback through R_2, the equilibrium output level is made zero, and so a three-level output is now produced.

When the circuit is in the equilibrium state, a positive trigger pulse will raise the noninverting input voltage through D_1 and cause the output to swing positive. This output swing supplies positive feedback through C_1 and D_1 to hold this state as long as the positive feedback exceeds the negative feedback, with $i_2 > i_1$. If $R_4 < R_2$, the positive feedback is initially greater for the essentially equal voltages on the two resistors. As the voltage on C_1 increases, i_2 decreases until it equals i_1, for which the output falls toward zero. With this fall, C_2 couples positive feedback through D_2 to make the output continue its negative swing to its negative

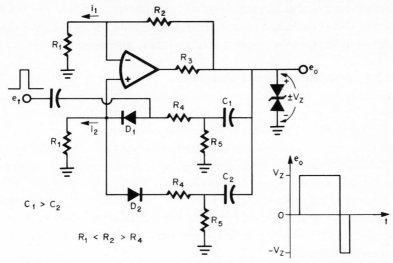

Fig. 5.21   Two sequential pulses are produced by monostable action where separate positive feedback paths are provided for the two output pulses.

state.   The output remains in this state until the positive feedback current declines to the level of the negative feedback current as C_2 discharges.   At this point the output rises and would couple positive feedback through C_1 again if C_1 were not chosen to be greater than C_2. Under this condition the voltage on C_1 will not change enough during the shorter C_2 timing cycle to forward-bias D_1 when the output reaches zero.   As a result, the output rise stops at zero until the circuit is triggered again.   Before retriggering, the capacitors should be allowed time to recover their equilibrium zero voltage states.   With C_1 greater than C_2, the positive pulse must be the longer one, but the time intervals can otherwise be separately set.

5.5   Modulated Signal Generators
Signal generators which have a voltage control over some waveform characteristic produce waveforms which are modulated by the control voltage.   A variety of waveform characteristics are commonly modulated for information transmission, and operational amplifier signal generators can be precisely controlled for most of these modulation approaches.[1] Many of the previous signal generators in this chapter can be modulated, including those of Figs. 5.4, 5.5, 5.7, 5.8, 5.12, 5.16, and 5.17.   The

modulation of these signal generators is described in this section along with other modulated signal generators. With these circuits modulation is achieved over amplitude, frequency, and pulse width.

5.5.1 Amplitude-modulated generators   To modulate the amplitude of virtually any signal, the signal can be multiplied by the modulating signal. For the general case an analog multiplier is the most straightforward modulator, although some simpler circuits can be used to modulate the amplitudes of square waves and pulse trains. Through the use of multipliers the quadrature oscillator presented in Fig. 5.7 computed a feedback signal to stabilize amplitude. While providing the required amplitude stabilization, this feedback circuit also permits amplitude modulation by the control voltage. If a modulation signal e_m is applied to the control input, the sine wave generated will have a modulated amplitude of

$$A = \sqrt{-e_m}$$

Since the amplitude-control feedback is continuous, the response to the modulating signal is as fast as the multiplier responses permit. The multipliers also determine the modulation range by their dc offset limitations on accuracy for small amplitudes.

Amplitude modulation of a square wave or a pulse train can be achieved without multipliers by using the signal to control the amplification of a modulation signal. Specifically, the square wave or pulse is used to control gain polarity. When a signal causes this gain polarity to alternate, the amplified signal alternates in polarity, tracing out a modulated signal as illustrated in Fig. 5.22. The modulated square wave shown is produced as the square-wave generator formed with A_1 switches the gain polarity of the amplifier formed with A_2. Operation of the square-wave generator was described in Sec. 5.2.1, and it has a highly predictable oscillation frequency, as long as the differential input resistance of the amplifier remains high with large differential input voltages.

Square-wave switching of the amplifier formed with A_2 reverses the amplifier gain polarity by converting A_2 from an inverter to a voltage follower. This conversion is like that previously performed for absolute-value conversion (Fig. 4.14). When the output of A_1 is negative, Q_1 is ON to ground the noninverting input of A_2. The ON resistance of the switch produces an input error voltage, but this error is compensated by

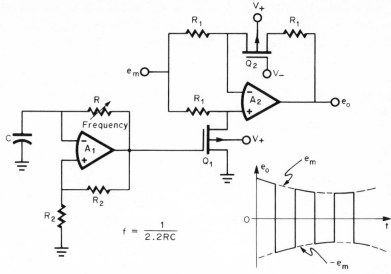

Fig. 5.22 An amplitude-modulated square wave is generated when a square wave is used to alternate the polarity of the gain by which a modulation signal is amplified.

$$f = \frac{1}{2.2RC}$$

connecting a matching ON resistance in the amplifier feedback. As biased, the compensating FET Q_2 acts simply as a resistor. With Q_1 ON, A_2 appears as a unity-gain inverter to the modulation signal e_m. When Q_1 is OFF with the positive state of A_1, the modulation signal reaches the noninverting input of A_2 directly. No voltage is then across the R_1 summing resistor, and so essentially no feedback current flows around A_2. With no feedback current there is no feedback voltage drop, and the output of A_2 follows its input signal e_m. In other words, A_2 has been switched to a voltage follower from an inverter to reverse the polarity of the gain it provides to e_m.

5.5.2 Frequency-modulated generators Frequency-modulated sine waves, triangle waves, square waves, ramp trains, and pulse trains can be generated by operational amplifier signal generators. For frequency-modulated sine waves, the circuits of Figs. 5.4 and 5.5 can be voltage-controlled. While the phase-shift oscillator of Fig. 5.4 is resistor-controlled rather than directly controlled by a voltage, it is readily adapted for frequency shifting by a control signal. To shift frequency one or both of the phase-shifting resistors, previously shown as R_1 and R_2, are shunted by other resistors. By electronically switching these shunt resistors, the frequency shift can be voltage-controlled for virtually any number of steps in frequency. Both phase-shifting resistors should be shunted for wide-frequency-range operation so that both phase shifters have significant control at each frequency.

To modulate the frequency of a sine wave in direct proportion to a modulation voltage, the multiplier-controlled quadrature oscillator presented in Fig. 5.5 can be driven by a modulation signal. This results in a modulation described by

$$f = \frac{e_m}{20\pi RC}$$

where e_m is a modulating signal of much lower frequency. A highly linear modulation over a wide range of frequencies can be achieved with this circuit, as it is controlled primarily by the linearities and frequency responses of the multipliers. Some means of amplitude control must be added to the oscillator. Fairly simple amplitude stabilization can be applied where the only important waveform characteristic is frequency. If a highly stable amplitude over a wide frequency range is required, the computing feedback shown in Fig. 5.7 can be used.

Frequency-modulated ramp and pulse trains are somewhat easier to develop and can be generated with the circuits of Figs. 5.16 and 5.17. In both circuits the signal frequencies are determined by a voltage input to an integrator. Previously the negative supply voltage served as the integrator input voltage, but this can be replaced by a modulation signal e_m for direct, linear control over frequency. With the single amplifier circuit of Fig. 5.16, the ramp and pulse trains are then frequency-controlled by e_m, as expressed by

$$f = \frac{e_m}{V_P RC}$$

Both circuits offer wide modulation frequency ranges, although the frequency sensitivity of the clamp of Fig. 5.16 makes this circuit more limited. Otherwise, the frequency ranges are limited by the integrator drifts from amplifier dc input errors[1] and by the amplifier slewing rates.

Frequency-modulated square waves and triangle waves can also be generated by modulating integrator input voltages. While it was possible to directly connect the modulating signal to the integrator in the ramp and pulse generators above, the polarity of the integrator input must be reversed each half cycle for the square- and triangle-wave generators to reverse the direction of integration. This operation is achieved directly with a control multiplier in the circuit of Fig. 5.12. By inserting a multiplier in the basic integrator/comparator feedback loop, the alternating comparator output is multiplied by a control voltage. If this control voltage is made the modulating signal e_m, the oscillation frequency will be

$$f = \frac{e_m}{20RC}$$

The modulation linearity and range will be essentially set by the linearity of the multiplier, the drift with time of the integrator, and the responses of the feedback loop.

As an alternative to the multiplier control above, a frequency-modulated triangle- and square-wave generator can be formed with the aid of the square-wave amplitude modulator from Fig. 5.22. Just as with the multiplier control above, modulation of the amplitude of a square-wave input to an integrator varies the integration time between comparator trip points. By this control over integration time, a frequency modulation is attained with the integrator/comparator configuration, as shown in Fig. 5.23. The basic generator consists of integrator A_2 and comparator A_3, with operation as described for Fig. 5.11. Added to the normal

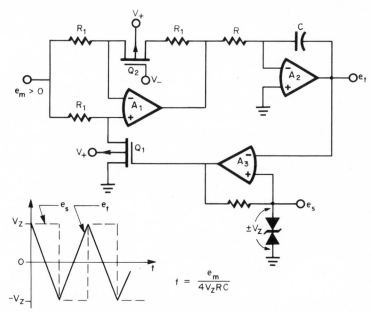

Fig. 5.23 Frequency-modulated square and triangle waves are generated by the circuit of Fig. 5.11 when it is combined with the amplitude modulator of Fig. 5.22.

feedback loop is the modulator formed with A_1. The polarity of the gain provided by A_1 to e_m reverses each time the integrator output reaches a comparator trip point and causes the comparator to reverse the state of switch Q_1. As described in Sec. 5.5.1, reversing the state of this switch converts the amplifier configuration from that of an inverter to a follower, and this reverses the polarity of the modulation signal reaching the integrator input. With either gain polarity, e_m controls the magnitude of the integrator input voltage and thereby controls frequency by the relationship

$$f = \frac{e_m}{4V_ZRC}$$

5.5.3   Pulse width-modulated generators

Simple circuitry and high power efficiency characterize pulse width modulation. Since the pulse is generated by switching, control is simplified, and the modulator output is always either saturated or off for low power loss. The power efficiency is especially advantageous in the controllers of the next chapter where higher output power is needed. Modulation control of the switching fixes the relative time intervals of the two output pulse states, as described with the circuits below.

Highly precise pulse width modulation is attained with a triangle-wave generator and a comparator, as illustrated in Fig. 5.24. The triangle-wave generator can be formed as described in Sec. 5.2. If the triangle wave is compared with the level of a modulation signal, a positive output pulse is generated when the triangle wave is greater. When the modulating signal is zero, the triangle wave is greater one-half the time, so that the generated pulse then has a duty cycle of one-half. As the modulating signal increases, it linearly decreases the interval of the positive pulse, since the triangle wave is a linear function of time. The result is an output pulse train with an average value that is directly proportional to the modulating signal and with a duty cycle of $0.5 - e_m/2E_p$, where E_p is the amplitude of the triangle wave. The linearity of this modulation is determined by that of the triangle wave. Any comparator switching delay will limit the extremes of the duty cycle range, but otherwise equal positive and negative switching times will have compensating errors.

By using the integrator/comparator configuration for a triangle-wave generator, linear pulse width modulation can be produced without a

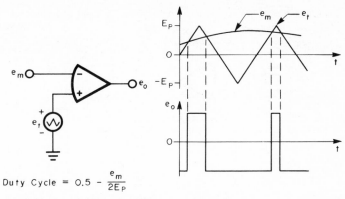

$$\text{Duty Cycle} = 0.5 - \frac{e_m}{2E_P}$$

Fig. 5.24 By comparing a modulation signal with a triangle-wave reference, a linear pulse width modulation is achieved.

separate comparator. With this configuration, presented in Fig. 5.11, the modulating signal can be summed into the integrator input on a resistor. During a given cycle, the addition of this modulation signal will increase the integration rate in one direction and decrease it in the other direction. The result is an unsymmetrical triangle wave and a corresponding modulated pulse output from the comparator. With equal integrator summing resistors, the duty cycle of the output pulse is equal to $0.5 - e_m/2V_Z$, provided the symmetry control of that circuit is set for $S = 1$. Alternatively, the symmetry control can be replaced with the modulation signal for a high impedance modulation input.

Although it is nonlinear, a pulse width modulation can be produced by summing a modulation signal into the single amplifier triangle-wave generator. For feedback controllers the modulation nonlinearity is overcome by the feedback, and this simple approach need not limit precision. Shown in Fig. 5.25 is the result of this summation in the basic circuit from Fig. 5.8. The effect of the modulation signal is analogous to that just described above in that one integration rate is increased and the other is decreased to alter the duty cycle. Since the capacitor charging is nonlinear, the effect of e_m is nonlinear as expressed by the duty cycle of

$$\text{Duty cycle} = \frac{\ln(1 + 2V_Z/e_m)}{\ln(1 - 4V_Z^2/e_m^2)}$$

As always with this basic square- and triangle-wave generator, the operational amplifier input impedance must remain high under overload so that amplifier input currents do not affect capacitor charging.

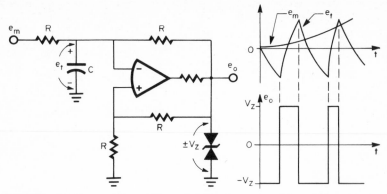

Fig. 5.25   Nonlinear pulse width modulation is produced by adding charging current from a modulation signal to the elementary square- and triangle-wave generator.

REFERENCES

1. G. E. Tobey, J. G. Graeme, and L. P. Huelsman, *Operational Amplifiers: Design and Applifications,* McGraw-Hill Book Company, New York, 1971.
2. T. Cate and H. Handler, Designing with Packaged Analog Multipliers, *EEE,* May 1969.
3. J. G. Graeme, Triangle-Wave Circuit Has Wide Range Controls, *EDN/EEE,* April 1972.
4. W. Neeland, Function Generator Has Variable Polarity Exponents, *Electron. Eng.,* December 1969.
5. J. G. Graeme, Pulse Generator Offers Wide Range of Duty Cycles, *EDN/EEE,* Sept. 1, 1971.
6. J. G. Graeme, Simple Op Amp Relaxation Oscillator Generates Linear Ramp Output, *EDN/EEE,* March 1, 1972.
7. J. G. Graeme, Op Amps Generate Precision Staircase, *Electronics,* Jan. 31, 1972.
8. P. Weil, Feedback and Clamping Circuits Improve Comparator One-Shot, *Electron. Design,* April 12, 1970.
9. G. Flynn, Sequential Bipolar Multivibrator, *EEE,* April 1969.

6

SPECIAL-PURPOSE CIRCUITS

In the previous chapters general-purpose operational amplifier circuits were described. Numerous special-purpose circuits of general interest are presented in this chapter, including transistor test circuits, measurement circuits, controllers, and others. For transistor testing, operational amplifiers provide simple, independent control of dc biases and test signals to test virtually any bipolar transistor or FET characteristic. Operational amplifiers permit simple circuit configurations for precise meter circuits and for circuits that measure frequency, phase, and high voltage. High gain in operational amplifiers produces a sensitive monitor of system deviation from a desired setpoint in controller applications for on-off, proportional, and optimizing control. Other special-purpose circuits described here include some for audio and AGC.

6.1 Transistor Test Circuits

All the basic low-frequency characteristics of bipolar transistors and FETs can be measured with circuits using only one or two operational amplifiers.[1]   These circuits are described, with comments on accuracies and alternate techniques, below.   Only resistors are adjusted in the test circuits, as will be described, to provide independent adjustments of dc bias current, dc bias voltage, and the ac test signal.   Resistive controls without interaction are made possible by the high-gain feedback, high input resistance, and low output resistance provided by an operational amplifier.   While the basic test circuits are intended for voltmeter display, they can be used with a simple ramp generator to provide oscilloscope traces of a wide variety of transistor parameters.

6.1.1   Gain measurement

Curve tracer bias of bipolar transistors sets the base current for a common-emitter configuration.   Beta is then found by observing the resulting collector current.   This is a poor bias approach for beta testing, as the magnitudes of the collector bias and signal currents then vary with beta.   Instead, it is better to measure beta at a specific collector bias current and signal current.   Beta is then determined from the resulting base current magnitude.   Fixed collector bias and signal currents are established in the beta test circuit of Fig. 6.1 by the feedback around A_1.   In establishing these currents, A_1 drives the transistor to maintain near-zero input voltage and current at the amplifier inputs.   For zero input voltage, the common ends of R_1 and R_2 are at ground level,

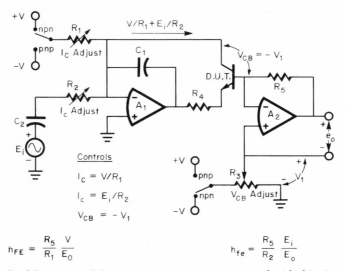

$$h_{FE} = \frac{R_5}{R_1} \frac{V}{E_o} \qquad\qquad h_{fe} = \frac{R_5}{R_2} \frac{E_i}{E_o}$$

Fig. 6.1   Static and dynamic current gains are measured with this circuit at bias and signal levels set by independent control potentiometers.

and the entire input signals are impressed on these resistors. For zero amplifier input current, the currents from R_1 and R_2 all flow in the collector. The dc and ac collector currents are then $I_C = V/R_1$ and $I_c = E_i/R_2$, respectively.

The base current resulting from the above collector currents is supplied to A_2 and flows in the feedback resistor R_5. From this current an output voltage is generated equal to

$$e_o = \left(\frac{I_C}{h_{FE}} + \frac{I_c}{h_{fe}}\right) R_5$$

where h_{FE} is the dc or static value for beta and h_{fe} is the ac or dynamic value. Using the last result, the two current gains are expressed in terms of the dc and ac components of e_o, E_O and E_o, by

$$h_{FE} = \frac{R_5}{R_1} \frac{V}{E_O} \qquad h_{fe} = \frac{R_5}{R_2} \frac{E_i}{E_o}$$

Amplifier A_2 also serves as a convenient means of setting the collector-base bias V_{CB}. From R_3 a voltage is set at one amplifier input and is, thereby, set at the test transistor base. This base voltage establishes V_{CB}, since the collector is held at zero voltage by A_1. Then, from Fig. 6.1, $V_{CB} = -V_1$.

Using this circuit, low-frequency measurements can be made within the accuracy permitted by the temperature coefficient of beta. This temperature coefficient is about $0.5\%/\degree$C near room temperature, and normal ambient variations limit beta test accuracy to around 1 percent. Test accuracy is also directly related to the accuracies of voltages V and E_i and of resistors R_1, R_2, and R_5. At low currents an additional error in the test for h_{FE} is the input bias current of A_2, but this may be avoided by using an FET input operational amplifier. Further limitations on the h_{fe} test are the amplifier bandwidth and stray capacitance shunting the high resistance of R_5. These latter factors limit h_{fe} testing to 10 kHz or lower. By using a low test frequency with low-level tests, the stray capacitance effects which often limit curve tracer measurement by hysteresis looping are avoided.

The forward transconductance g_{fs} of a field-effect transistor (FET) can be measured in a manner similar to that of the beta test above. Again it is preferable to test at a fixed operating point rather than with a fixed driving signal. Therefore, the drain bias and signal currents are fixed in Fig. 6.2, as were the collector currents in Fig. 6.1. In this case, the dc

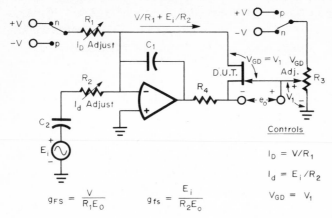

Fig. 6.2 Static and dynamic forward transconductances are measured in this circuit as bias and signal levels set by independent control potentiometers.

and ac drain currents are set by R_1 and R_2 to be $I_D = V/R_1$ and $I_d = E_i/R_2$, respectively.   The biasing voltage on the FET is set directly by establishing the gate voltage with R_3.   Since the drain is held at zero voltage at the operational amplifier input, the gate-drain voltage V_{GD} is set by the gate bias at $V_{GD} = V_1$.   To find g_{fs}, the gate-source voltage v_{gs} developed by the drain current is measured at the output.   By relating this voltage to the current, the static and dynamic values of forward transconductance are found in terms of the dc and ac components of e_o, E_O and E_o, as

$$g_{FS} = \frac{V}{R_1 E_O} \qquad g_{fs} = \frac{E_i}{R_2 E_o}$$

As before, the primary limitation on test accuracy is generally the temperature coefficient of the device under test.   The exact coefficient encountered varies from zero to $0.7\%/°C$, depending upon how close the test current is to the FET zero temperature coefficient point.   The circuit-oriented accuracy limitations are the same as those encountered with the corresponding elements in the beta test above.   Test bandwidth is again dc to 10 kHz.

6.1.2 Breakdown and leakage testing   The breakdown voltage of a transistor is not an exact voltage, since this voltage varies in several ways with the current flow under breakdown. As this breakdown current is the source of circuit error, it is desirable to set the current level at which breakdown voltage is measured. Such testing is readily performed with an operational amplifier, as illustrated in Fig. 6.3 for several cases. The amplifier output increases the voltage on the device under test until the device breaks down and supplies a feedback current to the amplifier input. With this feedback, the amplifier input voltage and current are returned to zero, and the breakdown current I_{BV} will be $I_{BV} = V_+/R$. Any breakdown voltage can be tested in this manner by connecting the appropriate device terminals between the amplifier input and output. The connection polarity is that which results in breakdown with a current flow into the terminal connected at the amplifier input.

Errors in this breakdown voltage are negligible for the measurement accuracies normally desired. Test accuracy is maintained at low current levels until the input bias current of the operational amplifier becomes significant. Since this limit can be quite low, breakdown testing at very low currents, for close examination of junction quality, is possible. However, the range of voltages accommodated in this test is more limited. It is limited by the output voltage range of the operational amplifier, and a high-voltage amplifier is required. Often breakdown voltages that exceed the capabilities of most high-voltage operational amplifiers are encountered. In this case the output voltage range can be extended, as

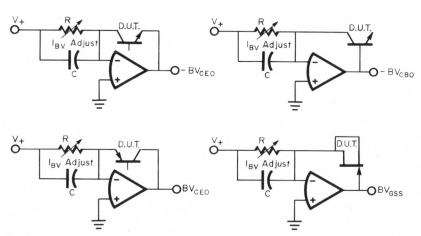

Fig. 6.3 Direct reading of breakdown voltage with independent current adjustment is provided by an operational amplifier.

shown in Fig. 6.4. The circuit shown adds voltage gain following the amplifier with a common-emitter p-n-p driving an FET current source load. By using the FET, a high gain and a current-limited output are achieved simply. Output current drawn through the test device is limited to the I_{DSS} of the FET. Further voltage-range improvement can be achieved with a stacked output stage.[2] By stacking output transistors in series, the voltage range can be made greater than the individual breakdown levels of these transistors.

The very low leakage currents of silicon bipolar transistors and FETs present a significant measurement challenge. Bipolar transistor leakages of 0.1 nA and FET leakages of 0.1 pA are now attained. Such low current levels cannot be measured on most curve tracers or specialized transistor testers. Fortunately, an operational amplifier readily permits measurement of these low-level leakage currents, as demonstrated in Fig. 6.5. In each case the leakage current flows in the amplifier feedback to develop an output voltage directly proportional to the leakage current. Feedback constrains one terminal of the device under test to zero voltage at the amplifier input, and the bias voltage is established by a potentiometer which controls the voltage at the other device terminal.

The principal limitations on test accuracy and test current range are the input bias current of the amplifier, stray leakages, and noise. To avoid input bias current error, which adds directly to the leakage current, an operational amplifier is chosen which has negligible input bias current. For most bipolar transistor leakage current testing, an FET input

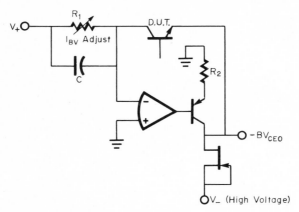

Fig. 6.4 Breakdown voltages exceeding the operational amplifier output voltage range can be tested by adding a gain stage to the amplifier.

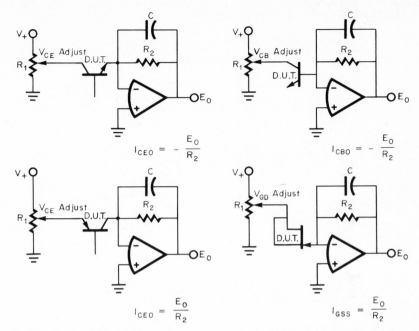

Fig. 6.5 Extremely low leakage currents are measured accurately by using FET or varactor input amplifiers which have comparatively negligible input bias currents.

amplifier has negligible input current. For FET leakage current measurement, a varactor input operational amplifier and a very high feedback resistance are required. For all cases stray leakages should be minimized by the use of clean surfaces and isolation from supply or bias voltages. To reduce noise, various degrees of shielding and filtering are needed. In general, bypass of the feedback resistor provides adequate filtering.

6.1.3 Output resistance measurement

For high gain in common-emitter amplifiers, high resistance loads are used. The ability of a bipolar transistor to drive a high resistance load is expressed by its output conductance h_{oe}. To measure h_{oe}, the slope of the common-emitter characteristic curve can be calculated with a curve tracer. Alternatively, this small slope can be measured accurately and quickly with the circuit of Fig. 6.6. Basically, this test circuit provides an output voltage proportional to the emitter current change induced by a collector-emitter voltage change. The collector-emitter signal voltage is established by E_1, the emitter signal current is supplied by E_o from A_2, and biasing is set by A_1

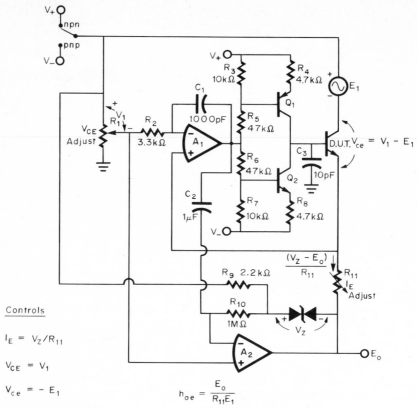

Fig. 6.6 Output conductance measurement is performed by this circuit with biases set by R_1 and R_{11}.

and A_2.   A differential current source formed with Q_1 and Q_2 and controlled by A_1 biases the base of the test transistor.   With current source base bias, the biasing impedance will be much greater than the load impedance presented by the transistor.   Thus, the output conductance measured will be the worst-case value h_{oe}.   Operational amplifier feedback around the current sources from the test transistor emitter forces the emitter voltage to equal that presented by R_1.   This sets the collector-emitter dc voltage V_{CE} at $V_{CE} = V_1$.

The other amplifier is also biased from R_1 so that the adjustment of V_{CE} does not affect the emitter bias current I_E.   As R_1 causes A_1 to vary the test transistor emitter voltage, this potentiometer also causes A_2 to identically vary the dc voltage at the other end of the emitter resistor R_{11}. No change in emitter resistor voltage results, and I_E is unaffected.   Instead, the dc voltage on R_{11} is established by the zener diode in the feedback around A_2. The zener drops the dc output voltage of A_2 below its input voltage, which is also the emitter voltage. Then $I_E = V_Z/R_{11}$.

If the output voltage of A_2 had a fixed dc level only, then emitter signal currents created by V_{ce} variation would not develop. Instead, the feedback provided by A_1 would drive the base to hold the emitter voltage constant and would, thereby, remove the current which is to be measured. The desired emitter signal is restored by the ac coupling from the output of A_1 to A_2. From the signal coupled, the output of A_2 drives R_{11} to supply the signal current and to present a buffered output for measurement. This yields a signal output of $E_o = I_e R_{11}$ and an h_{oe} given by the ratio I_e/V_{ce}, or

$$h_{oe} = \frac{E_o}{R_{11}E_1}$$

Other than errors in E_1 and R_{11}, the major accuracy limitations are the finite gain and bandwidth of A_2. To supply all the emitter signal current, A_2 must have a current gain through R_{11} that is much greater than the beta of the transistor tested. This commonly sets the upper limit of the test bandwidth at about 200 Hz. A lower bandwidth limit is similarly imposed by coupling capacitor C_2. Within this bandwidth it is possible to make low-frequency h_{oe} measurements which avoid the stray capacitance hysteresis looping encountered on many curve tracers at low current.

Somewhat simpler than h_{oe} testing is the measurement of the drain-source resistance r_{ds} of an FET. For an FET, its r_{ds} is the measure of its ability to drive high resistance loads. The r_{ds} test circuit of Fig. 6.7 is analogous in form to the previous h_{oe} test circuit, but a high driving impedance is not required for the FET gate. Once again, A_1 fixes the dc bias voltage at V_1, giving $V_{DS} = V_1$. The feedback zener sets the dc voltage on the source resistor, giving $I_S = V_Z/R_2$. As before, A_2 supplies the signal current, which is now $I_s = E_o/R_2$, and the drain-source resistance is found from

$$r_{ds} = \frac{E_1}{E_o} R_2$$

This result is accuracy-limited by errors in E_1 and R_2 and by the finite gain of A_2. In this circuit, the current gain through A_2 and R_2 must be much greater than the g_{fs} of the FET in order to have E_o supply all the

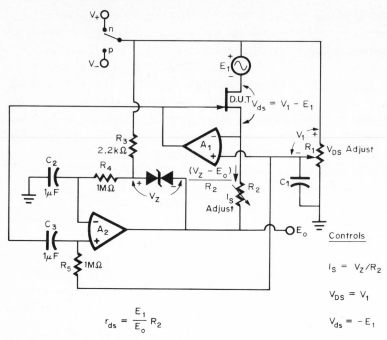

Fig. 6.7 At an operating point set by R_1 and R_2, the drain-source resistance of an FET is measured under a signal voltage E_1.

signal current.   Measurement bandwidth is, then, about 20 Hz to 20 kHz.

6.1.4 Noise measurement   To simplify noise considerations, the noise sources of transistors can be considered as equivalent noise generators at the terminal used for the signal input.   In this way the noise contributed by the transistor can be compared directly with the input signal to determine signal sensitivity.   This noise representation closely parallels that used for operational amplifiers.[2]   All transistor noise sources are considered at the appropriate terminal, and the combined effects are represented by an equivalent noise voltage generator in series with the terminal and a noise current generator that supplies current to the terminal. Measurement of these equivalent input noise sources is straightforward with feedback circuits.

For a common-emitter bipolar transistor or a common-source FET, the equivalent input noise signals can be measured with the circuit of Fig. 6.8.   When the transistor is connected in an operational amplifier feedback loop, as shown, the equivalent input noise sources are accurately related to output signals, and biasing is easily controlled. Biasing is fixed when feedback reduces the differential input voltage of the amplifier to near zero. Then $V_{CE} = V_1$, and the transistor operating point is set by R_3 and R_4. The transistor acts as an added gain stage for the amplifier, and overall feedback is supplied by R_1, R_2, and C. Since the transistor adds voltage and current gain, it reduces the equivalent input noise contributed by the operational amplifier input noise voltage and current. With reasonable care the reduced noise contribution of the amplifier can be made negligible in comparison with that of the transistor. Then

$$e_o \doteq \left(1 + \frac{R_2}{R_1}\right) e_{n1} \gg \left(\frac{e_{n2}}{R_3} + i_{n2}\right) r_e$$

where e_{n1} is the equivalent input noise voltage of the transistor and e_{n2} and i_{n2} are the operational amplifier input noises. In an analogous manner the equivalent input noise voltage of an FET is measured with this circuit.

For both cases, the added gain of the transistor may necessitate added phase compensation. The phase compensation will fix the overall amplifier bandwidth and, thereby, the maximum noise measurement bandwidth. Further restriction of the measurement bandwidth is readily

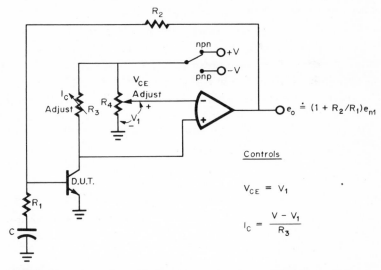

Fig. 6.8 To measure the equivalent input noise of a transistor, the transistor is inserted as an input stage in an operational amplifier feedback loop.

achieved with capacitive bypass of R_2. A low-frequency limit is established by R_1 and C at their break frequency. Whenever the measurement bandwidth includes frequencies of the noise environment, careful
shielding must be used.

To measure the equivalent input noise current of a transistor, the circuit of Fig. 6.8 is modified to amplify this noise signal rather than the
noise voltage. This modification deletes R_1 and C to reduce the voltage
gain to unity and increases R_2 so that it develops a measurable voltage
from the input noise current. The level of R_2 required is that which
makes the output signal created by noise current much greater than the
equivalent noise voltage. Then, if the noise current of the amplifier is
adequately reduced by the transistor beta,

$$e_o \doteq i_{n1} R_2 \gg \frac{i_{n2}}{\beta} R_2 + e_{n1}$$

where i_{n2} is the input noise current of the operational amplifier. For a
bipolar transistor this measurement can be made highly accurate with
careful shielding. However, the extremely low noise current of FETs
can be measured only with exceptionally high resistance levels and the
greatest attention to shielding.

6.1.5 Other transistor test circuits   When operational amplifiers are
used to provide control of biases and test signals, several other transistor
characteristics can be measured. Described below are test circuits for
pinchoff voltage measurement, transistor matching, and oscilloscope display of characteristics. In order to easily measure the pinchoff voltage
V_P of an FET, the current in the FET must be nearly pinched off. The
low current level remaining is sensitive to voltmeter loading, but can be

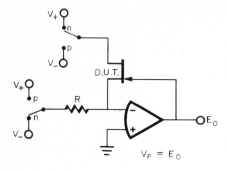

Fig. 6.9 Pinchoff voltage measurement is undisturbed by voltmeter loading with the buffered
output of the biasing operational
amplifier.

buffered with an operational amplifier which also biases the FET. Figure 6.9 shows such a circuit. With the FET source voltage driven to zero at the amplifier input, the output voltage is V_{GS}. This voltage approximates V_P for a specified, small source current supplied through the resistor R.

Matched pairs of bipolar transistors or FETs are essential for differential amplifier stages with low dc error and thermal sensitivity. Such transistor pairs have a match in emitter-base voltage V_{BE} or a match in gate-source voltage V_{GS}. Elementary biasing is not generally suitable for matching purposes, as a mismatch, ΔV_{BE} or ΔV_{GS}, creates a compensating current unbalance between the two transistors. Operational amplifiers can remove this interdependence. In Fig. 6.10 a matching circuit is shown which makes the current balance between the transistors independent of their emitter-base voltage difference. The individual bias currents and voltages are also independently controlled. Although shown with bipolar transistors, the same circuit provides V_{GS} match-testing for FETs. To set the two collector-emitter bias voltages, R_1 sets the voltage at the inputs of both amplifiers and, thereby, at the two emitters. This gives $V_{CE} = V_1$. With the two emitters at the same voltage, the base voltages will differ by ΔV_{BE}. This mismatch is measured as $E_O = \Delta V_{BE}$.

For the above output voltage to represent the transistor mismatch, the emitter currents must be identical. Feedback sets the two emitter currents as the amplifiers drive the transistors to accept the currents supplied by the resistors labeled R_2. These resistors carry emitter currents derived from the zener diode voltage V_Z. Since the diodes are referenced to one amplifier output, rather than to ground, a change in V_1 from R_1 does not alter the voltage on the current-setting resistors R_2 and R_3. The voltage across R_2 and R_3 is $-V_{BE} + V_Z + V_f$. Since the diode voltage V_f will approximately cancel $-V_{BE}$, the drop on R_2 and R_3 is V_Z. The emitter currents are then

$$I_E = \frac{V_Z}{R_2 + 2R_3}$$

The dominant measurement errors are those created by differences in operational amplifier offset voltages, in the emitter resistors R_2, and in the temperature of the two transistors. A difference in input offset voltages ΔV_{OS} results in an identical difference in the transistor emitter volt-

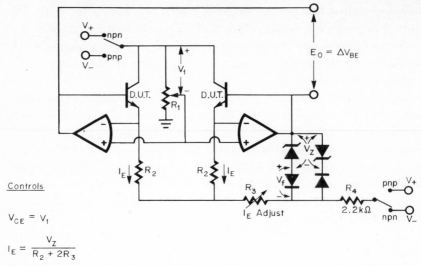

Fig. 6.10 Measurement of V_{BE} match, or similarly V_{GS} match, is made with accurately matched currents set by the same bias source.

ages to directly produce an output error equal to ΔV_{OS}. To remove this error, ΔV_{OS} is nulled by adjusting for $E_O = 0$ when the emitter-base junctions are shorted.

A difference between the resistors labeled R_2 creates a current unbalance and an associated error $\Delta V'_{BE}$ of

$$\Delta V'_{BE} = 25 \text{ mV} \ln \frac{I_E}{I_E + \Delta I_E}$$

Since the currents in this expression are directly determined by the resistors, the expression may be rewritten in terms of the resistances as

$$\Delta V'_{BE} = 25 \text{ mV} \ln \frac{R_2 + \Delta R_2}{R_2}$$

For small unbalances, the above resistance ratio is near unity, and the logarithm may be expanded to yield

$$\Delta V'_{BE} \doteq 25 \text{ mV} \left(\frac{\Delta R_2}{R_2}\right) \qquad \text{for } \Delta R_2 \ll R_2$$

This result indicates that only a 1 percent emitter resistor mismatch will produce a $\Delta V'_{BE}$ error of 0.25 mV. Removal of this error is accomplished by adjusting the R_2 resistors until interchanging the two transistors creates only a reversal in the polarity of E_O.

Another $\Delta V'_{BE}$ error is produced by the transistor temperature difference ΔT. From the $-2mV/^\circ C$ thermal sensitivity of a silicon junction voltage, the resulting error is

$$\Delta V'_{BE} = -(2 \text{ mV}/^\circ C)\Delta T$$

Just a $0.1 \,^\circ C$ difference results in a 0.2-mV error, and so thermal conditions must be very well matched. Matching of the thermal environments of the two transistors is improved by using adjacent test sockets and by avoiding air drafts and nearby thermal radiation. If each of the above error sources is controlled, overall measurement error may be reduced to 0.1 mV.

A wide variety of transistor parameter dependencies can be displayed on a common oscilloscope using the preceding circuits with a ramp generator. The ramp provides a linear sweep of one parameter so that the dependence of another characteristic can be observed, as with a curve tracer, but for a greater variety of measurements. In this way, i_s versus v_{gs} can be displayed to locate an FET zero temperature coefficient point. Or beta versus i_c can be viewed to determine the usable current range of a transistor. Similarly, g_{fs} versus i_s can be displayed for an FET. More conventional traces, such as the current-voltage characteristic of a junction, are also possible.

One example of this display technique is illustrated in Fig. 6.11. Here the i_s versus v_{gs} characteristic is produced for oscilloscope display. As with the pinchoff measurement circuit of Fig. 6.9, the output voltage

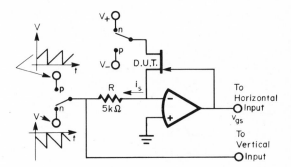

Fig. 6.11 A ramp generator drive replacing a bias voltage permits oscilloscope display of characteristic curves, such as for i_s versus v_{gs}.

equals v_{gs}. This voltage is displayed with respect to the ramp voltage that sets the source current. Display error common to this type of application is largely that created by nonlinearity in the sweeping ramp. However, a nonlinearity of 0.25 percent is acceptable, and such a ramp is easily produced with the circuits of Sec. 5.3.

6.2   Measurement Circuits

In addition to the transistor parameter measurements of the last section, numerous other measurements can be made with operational amplifier circuits. Operational amplifiers generally permit simple circuit configurations for precise measurements. Described first in this section are circuits for a voltmeter and an ammeter. Then, circuits are presented for measurement of frequency, phase, kilovolt-level voltages, and threshold voltages. Other circuits useful for measurement include the absolute-value circuits of Sec. 4.3, the peak detectors of Sec. 4.4, and the rms-to-dc converters of Sec. 4.5.

6.2.1   Meter circuits   By using operational amplifiers, the current through a meter can be made precisely dependent upon the absolute value of a signal. The normal current errors created by voltage drops on the meter and on rectifying diodes are avoided by connecting the meter circuit in the feedback loop of an operational amplifier.[3]   For use as a voltmeter, the meter is connected as in Fig. 6.12. Since feedback maintains near-zero voltage between the amplifier inputs, it forces a current of e_i/R_1 to flow in R_1 to develop a voltage matching e_i. Little current is drawn by the amplifier input, and so essentially the same current is sup-

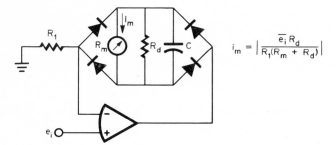

$$i_m = \left| \frac{\overline{e_i} R_d}{R_1(R_m + R_d)} \right|$$

Fig. 6.12   A voltmeter with precisely controlled meter current is formed by connecting the meter circuit in the feedback loop of an operational amplifier.

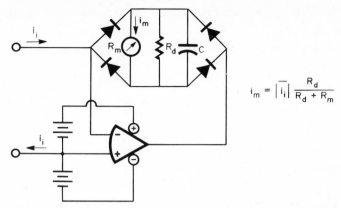

$$i_m = \left| \overline{i_i} \right| \frac{R_d}{R_d + R_m}$$

Fig. 6.13 By connecting the meter circuit in an operational amplifier feedback loop, an ammeter is formed which has a voltage drop that is only the differential input voltage of the amplifier.

plied through the meter circuit independent of the meter and diode voltage drops. The current is rectified by the diode bridge for a unipolar meter current. Meter damping is provided by R_d and filtering by C and R_d along with the meter resistance R_m for a time constant of $(R_d \parallel R_m)$ C. In this way, the average meter current is controlled to be

$$i_m = \left| \frac{\overline{e_i}}{R_1} \right| \frac{R_d}{R_d + R_m}$$

Only the small input bias current and input offset voltage of the amplifier limit the dc accuracy of this voltmeter beyond the limitations of the meter itself. The effects of the voltage drops on the meter, the diodes, and the damping resistor are reduced by a factor equal to the gain of the amplifier.

To form a low impedance ammeter, the meter circuit is again connected in the feedback loop of an operational amplifier, as shown in Fig. 6.13. Once again, feedback will supply that current needed to maintain near-zero voltage between the amplifier inputs. In this case the voltage so controlled is the drop across the ammeter. For this zero-voltage condition, the ammeter input current i_i must flow into the feedback meter circuit rather than into the amplifier input. Except for the small error of the amplifier input bias current, the meter current will be

$$i_m = \left| \overline{i_i} \right| \frac{R_d}{R_d + R_m}$$

With this current, voltage drops are developed on the meter resistance R_m, damping resistance R_d, and amplifier output resistance R_O, but the resulting voltage drop reaching the amplifier input is reduced by the amplifier gain A. Thus, the voltage across the ammeter due to the meter resistance is reduced by A, and the effective ammeter resistance is

$$R = \frac{R_m \parallel R_d + R_O}{A}$$

Added to the voltage produced with this ammeter resistance is the input offset voltage of the operational amplifier.

For general-purpose use an ammeter must be capable of being inserted in series with a current path. This requires that the ammeter return from one terminal the same current that it accepted into its other terminal. Such an operation is achieved with the above circuit through the use of the floating power supplies shown. The input current i_i is returned through the supply common, as it is the only current drain from the amplifier that does not flow from one supply through the amplifier to the other supply. Instead the meter current flows from one power supply and out the output terminal so that the supply currents are unbalanced by precisely i_i. This difference current flows from the supply common. Of course, if the amplifier has an internal bias return to common, the above condition does not hold.

6.2.2 Frequency and phase measurement circuits
To measure signal frequency or phase, a dc voltage that is proportional to either signal characteristic can be developed. This dc voltage is readily measured for the frequency or phase indication with a voltmeter such as that shown in Fig. 6.12. For frequency measurement the frequency-to-dc converter of Fig. 6.14 produces the desired output voltage by time-averaging the pulses developed with each signal cycle. Amplifier A_1 operates as a comparator to convert the signal to a controlled-amplitude square wave. If the input signal has unequal time intervals above and below the zero level, the comparator output will be similarly unsymmetrical, but this will not affect the output voltage developed. To improve the rise and fall times of an operational amplifier used for the comparator, its phase compensation should be removed.

Following the comparator, the signal is differentiated, rectified, and then averaged to produce a dc output. The differentiation and rectification are performed with A_2. From the rapid rises and falls of the com-

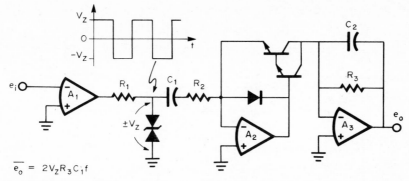

$$\overline{e}_o = 2V_zR_3C_1f$$

Fig. 6.14 Frequency-to-dc conversion is performed by deriving one current pulse per cycle and time-averaging the resulting pulse train.

parator output voltage, differentiation capacitor C_1 produces current pulses that are supplied to either the transistors or the diode around A_2. Only one polarity of current pulse is conducted by the transistors to the output amplifier A_3. This rectification action results in only one current pulse per cycle to A_3 rather than pulses with both the rise and fall of the signal. As a result, the time average of the pulses reaching A_3 is independent of the signal symmetry.

The averaging is performed by C_2, and the average current is determined by the change in capacitor charge that generated it and by the time between pulses. This is expressed by

$$\Delta Q = 2V_zC = \overline{i}\Delta t \qquad \text{for } R_2C_1 \ll \Delta t$$

Here the term Δt is the time between pulses, and this is the period of the signal or the inverse of the frequency. Thus, the average voltage generated by the flow of $\overline{i}$ in R_3 is related to the signal frequency by

$$\overline{e}_o = 2V_zR_3C_1f$$

Much of the ripple on this output is filtered out by the damping of any dc voltmeter used as a readout. At low frequencies the accuracy of this frequency-to-dc converter is determined primarily by the components in the above expression. However, at higher frequencies the parasitic and stray capacitances of the differentiator/rectifier circuit introduce charge errors. At some high frequency the slewing rate limit of A_1 will prevent it from completing its output swing, and very large errors will develop.

To measure the phase of a signal, its phase difference from a reference signal can be used to develop a proportional dc voltage. The circuit of

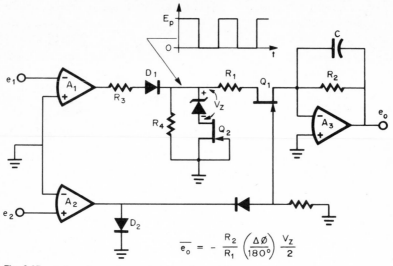

$$\overline{e_o} = -\frac{R_2}{R_1}\left(\frac{\Delta\varnothing}{180°}\right)\frac{V_z}{2}$$

Fig. 6.15 A dc voltage proportional to phase difference is developed by using the phase difference to control the time interval for which a nonzero signal is connected to an averaging amplifier.

Fig. 6.15 is such a phase-to-dc converter. It consists of two signal-squaring comparators A_1 and A_2 followed by an amplifier with a switched input. As before, operational amplifiers used for the comparators should not be phase-compensated, since this would reduce switching speeds. Comparator A_1 generates a zero-based square-wave input signal for the amplifier. So that the negative switching of this signal stops at zero, D_1 disconnects the amplifier, and R_4 discharges the capacitance of the zener diode. The square-wave signal is impressed on resistor R_1 and switch Q_1, so that the ON resistance r_{ON} of Q_1 affects the amplifier gain. To compensate for this effect, a matching ON resistance is added in series with the zener diode by means of Q_2 to increase the signal amplitude. As long as the current in the compensating Q_2 is less than I_{DSS}, the FET acts as a resistance. For accurate compensation the current in Q_2 is set to match that of Q_1 by appropriate selection of R_3. Then, the amplitude of the square wave will be

$$E_p = V_z\left(1 + \frac{r_{ON}}{R_1}\right)$$

The second comparator switches the input to the amplifier by driving

Q_1. If the two input signals are in phase, the amplifier input is connected to the square wave only when this signal is at zero, and so the amplifier output is zero. As a phase difference develops, the switch remains ON during a portion of the nonzero state of the square wave. This portion of the cycle is proportional to the phase difference, as is the time average of the signal reaching A_3. To average this signal a feedback capacitor is added, and the average of the output voltage becomes

$$\overline{e}_o = -\frac{R_2}{R_1}\left(\frac{\Delta\phi}{180°}\right)\frac{V_Z}{2}$$

where $\Delta\phi$ is the phase difference. Ripple remaining on this signal is filtered by most voltmeters that would be used as monitors. An offset is added to the output voltage if the switching times of Q_1 and the square wave differ. Addition of D_2 reduces this time difference by decreasing the voltage through which A_2 slews in turning OFF Q_1. The gate return-resistance of the FET switch must be low enough to rapidly discharge the FET capacitance during turn-on. Any dc offsets in the input signals must be removed to avoid errors in switching times.

6.2.3 Other measurement circuits
Among the other measurements performed with operational amplifier circuits are kilovolt-level voltage measurement and switching-threshold measurement. The high-voltage measurement is made with a low-voltage amplifier using charge amplifier techniques,[2] and threshold is measured using integrator feedback. To make the high-voltage measurement without loading errors, a small capacitor is charged to the high voltage, and then its charge is transferred to a much larger capacitor, where the charge develops a low voltage.[4] This operation is performed with the circuit of Fig. 6.16. First, C_1 is charged to the level of the high voltage through current-limiting resistor R. Then this charged capacitor is switched to the input of the charge amplifier, and again resistor R limits current flow. Without this limiting resistor the high voltage would be initially impressed on the amplifier rather than dropped across the resistor. Such a high-voltage input overload would probably destroy the amplifier, or any input protection clamp would dissipate the capacitor charge.

With the limiting resistor the discharge of C_1 is controlled so that virtually all the discharge current flows through C_2 to transfer the charge.

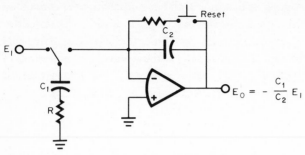

Fig. 6.16 To accurately measure a high voltage with a low-voltage meter, the voltage can be reduced with a charge amplifier.

This produces a low-voltage output from the amplifier that is an accurate measure of the high input voltage through the relationship

$$E_O = -\frac{C_1}{C_2} E_I$$

This reduced voltage can be measured with almost any voltmeter. Errors in this measurement technique are essentially due to the tolerances and dielectric losses of the capacitors. Virtually no measurement loading error is introduced, as the monitor capacitor C_1 draws only a leakage current when it reaches its final charge. Where such loading would not create significant error, a simple inverting amplifier connection with very low closed-loop gain can be used. Voltmeter loading on the holding capacitor C_2 is removed by the buffering of the operational amplifier. The amplifier loads C_2 with its input bias current, but this current can be made quite small by using an FET or varactor input amplifier.

To measure the switching thresholds of logic circuits and digital-to-analog converters, the input voltage is varied until the output just begins to switch. Integrator feedback around such switching circuits establishes a threshold-seeking operation for rapid testing. Such a feedback loop is represented in Fig. 6.17 for a D/A converter. If the converter output voltage differs from the reference voltage E_R, there will be a voltage across the integrator input resistor. Integration of this voltage difference produces a change in the converter input voltage E_I applied by the integrator. The integrator output continues to change the converter input voltage until the converter output equals E_R. Then the integrator capacitor holds E_I at this level. This equilibrium voltage will be the switching threshold if E_R is set near but not at the final value of the output voltage.

For the feedback connection shown it is assumed that there is no voltage inversion through the D/A converter. If the converter has complementary logic inputs, an inversion does occur, and the feedback and reference connections to the integrator must be interchanged to maintain negative feedback. The integrator time constant is chosen large enough to stabilize the feedback loop but not so large as to delay testing.

6.3   Controllers

Electronic analog controllers derive precision and stability from the high feedback-loop gain provided by operational amplifiers. The amplifiers provide precise comparison of reference and feedback signals for on-off control, and they produce continuous adjustment of a control signal in proportional controllers. For optimization control, operational amplifiers can generate a sweep signal that stops when a desired response is achieved. The various types of controllers described in this section can be used to control a variety of devices in addition to the heaters, motor, and receiver used for illustration.

6.3.1   On-off controllers   In the simplest applications a controller merely turns an actuator on and off as a system output falls below or rises above a reference setpoint. Such on-off controllers typically drive an actuator completely on and completely off, and this switched operation simplifies the actuator requirements. For switched operation the actuator control element can be a simple on-off device such as a relay, an SCR, or a solenoid valve. Hysteresis is often added to on-off control so that the actuator does not switch on and off too frequently. With hysteresis the actuator remains on for a system response somewhat above the setpoint

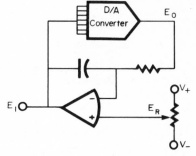

Fig.   6.17 Integrator   feedback around switching circuits automatically seeks out the switching threshold.

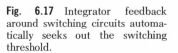

and then remains off as the response falls somewhat below the setpoint.

Both the setpoint comparison and the hysteresis are readily provided by an operational amplifier. Such control is achieved with the comparator connection shown for heater control in Fig. 6.18. Here a transistor switch is alternately turned on and off to control the power supplied to a heater element. The temperature of the object being heated is detected by a sensor, which is a diode in this case. Below $150\,^\circ$C, a forward-biased diode is an excellent temperature sensor due to its exceptionally linear voltage sensitivity to temperature[2] of around -2 mV/$^\circ$C. While the exact thermal sensitivity of a given diode is a function of its biasing current and its construction, its thermal variation is highly repeatable. Alternately, thermistors or thermocouples can be used as sensors in similar circuits, but the polarity of the associated thermal sensitivity must be considered to ensure that it supplies negative feedback to the comparator.

As temperature changes the sensor voltage V_f, the voltage is compared against the setpoint reference voltage E_R by the inputs of the operational amplifier. This comparison is altered by the positive feedback supplied through potentiometer R to develop hysteresis where desired. When the

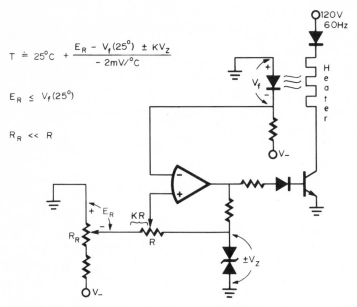

$$T \doteq 25\,^\circ C + \frac{E_R - V_f(25^\circ) \pm KV_z}{-2mV/^\circ C}$$

$$E_R \leq V_f(25^\circ)$$

$$R_R \ll R$$

Fig. 6.18   On-off heater control with accurate comparison of sensor and reference voltages and with hysteresis is provided by an operational amplifier comparator.

sensor voltage reaches that of the reference plus that of the hysteresis, the comparator switches to reverse the heater mode. This results in a controlled temperature of around

$$T \doteq 25°C + \frac{E_R - V_f(25°) \pm KV_Z}{-2 \text{ mV}/°C} \text{ for } E_R \leq V_f(25°) \text{ and } R_R \ll R$$

In practice, the precision of this control is limited by sensing accuracy. A time lag exists between a heater output and the sensing of its effect. While the heat travels to the sensor, additional heat is being generated that will produce overshoot. This time lag and the overshoot are reduced by placing the sensor close to the heater element, but the temperature sensed is then more that of the heater than that of the object heated. To sense the temperature of the object more accurately, some overshoot must generally be accepted with on-off heater controllers.

For reduced overshoot a multiple-level on-off controller or a proportional controller can be used. Proportional controllers are described later. An on-off controller with a multiple-level output decreases overshoot by reducing the actuator signal to a lower level when the system is near the setpoint. Away from the setpoint a full output is maintained for rapid response. To provide a two-level control, two comparators are used, as shown in Fig. 6.19 with a heater. Both comparators drive heater supply transistors, but the transistors supply different current levels. One transistor switch has a series-dropping resistor to limit its current supply to a lower level, and this lower output is supplied near the setpoint. Control of the lower level is provided by the comparator and hysteresis action of A_2 with the same operation as described for the previous circuit. However, the comparator action of A_1 is modified by the biasing of R_1 and R_2. With this added bias, A_1 switches before the setpoint is reached to turn off the unlimited, higher-level switch. By appropriate choice of the transition point and the reduced output level for a given system, overshoot can be greatly reduced without a similar reduction in response speed.

6.3.2 Proportional controllers Greatly improved control precision is achieved by developing a control signal that is proportional to the difference between the system state and the setpoint reference. Such proportional control gradually reduces the correction signal as the system approaches the setpoint to avoid most of the overshoot of on-off control. Sensing time lag still results in an excessive controller output, but the

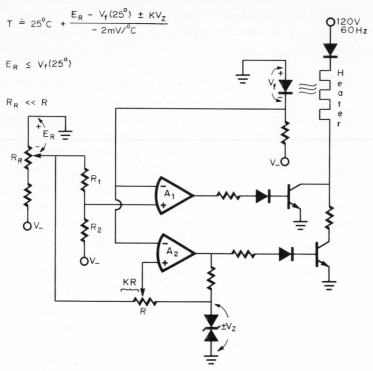

$$T \doteq 25°C + \frac{E_R - V_f(25°) \pm KV_Z}{-2mV/°C}$$

$$E_R \leq V_f(25°)$$

$$R_R \ll R$$

Fig. 6.19 With two comparators, an on-off controller supplies a reduced output near the setpoint to decrease overshoot.

output during the time lag is reduced so that the overshoot is smaller. Proportional control also supplies very small correction signals for fine adjustments.

In addition, a proportional controller can be made to supply any output level at equilibrium that may be needed to sustain an equilibrium state. Such a sustaining output is needed where there is a continuous drain at equilibrium on the element being controlled. Examples of this are the heat loss in high-temperature control and the leak loss in pressure control, which occur at a steady rate in equilibrium. In basic proportional control such losses are not compensated until they produce an error signal. If instead a continuous sustaining compensation is provided for these equilibrium losses, resolution is improved.

A sustaining proportional heater controller formed with two operational amplifiers is shown in Fig. 6.20. It is basically a pulse width modulation circuit with feedback from a heater to sensing diodes through the

medium being heated. If the sensor diodes change in temperature, their voltage drops change and produce a voltage difference from the reference level. This difference is integrated by A_1, and the resulting voltage change on R_2 alters the charging current to C_2. The charging and discharging rates of C_2 determine the output pulse width from A_2 as described in Sec. 5.5.3. As a result, the duty cycle of the heater current is altered, correcting the sensor temperature and returning the sensor voltage to that of the reference. As the sensor temperature is gradually corrected, the pulse width also changes due to the decreasing error. At equilibrium the pulse width settles at that width which sustains the zero error. This operation is made possible by the holding feature of the integrator. When the integrator input error signal goes to zero, the integrator output holds at the existing voltage level. This in turn holds the controller output width at that equilibrium level until a change in heat loss occurs.

Since the controller described is a null-seeking feedback system, nonlinearities in the controller are diminished in effect by the feedback-loop gain. This permits the use of a rather crude pulse width modulator when

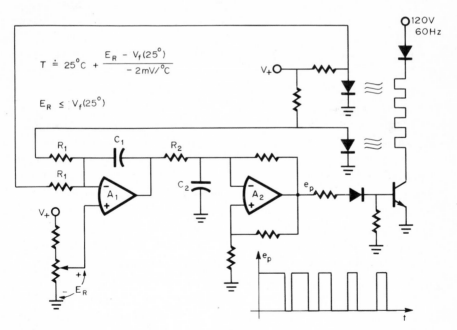

Fig. 6.20 An integrating error amplifier in this proportional controller holds the heater current pulse width at a sustaining value at equilibrium.

it is preceded by the high gain of an integrator. Resolution is then determined by the sensing and comparison accuracies. Sensing accuracy is primarily limited by sensor placement and the thermal transfer delays. Placement affects the thermal resistance and thermal loss between the object controlled and the sensors. Thermal transfer delays cause a hunting-type oscillation that is common to servo applications. To limit this hunting error, very long integrator time constants are used. This matches the controller response with the thermal response. Further improvement is provided by using multiple sensors with appropriately weighted integrator summing resistors. In this way, the controller response can be tailored to a weighted average of the thermal responses of many points on the object controlled.

For motor control a proportional controller can be made to supply a dither signal at equilibrium that avoids the delay and error introduced by motor-starting friction. In dc servo control a pulse width modulation commonly replaces a straightforward dc drive to improve power efficiency and to reduce power dissipation requirements on the controller elements. However, such a controller can be formed with two operational amplifiers to provide moderately precise dc servo position control.[5]

Unlike the elementary pulse width modulator which switches between two levels, the modulator required for dc servo control must switch between three levels. The servo motor drive voltage must switch from zero to a positive level for one rotational direction, and must switch from zero to a negative level for the opposite rotation. This three-level, or ternary, control is illustrated in Fig. 6.21a. Here the solid outlined waveform is the equilibrium condition, with dither pulses in both directions. Due to this dither signal, the motor shaft is in continuous motion, chattering about its null position. With this continuous motion, the motor-starting friction does not interfere to prevent small signals from moving the shaft or to create response delay. Rotation of the motor shaft is produced by increasing the widths of pulses of one polarity and decreasing the widths of the others. The dashed lines represent the pulse width increase which produces one direction of rotation. Both the positive and negative pulse widths must be alternately variable from zero to 100 percent.

Basically, this control is provided in the manner represented by the block diagram circuit of Fig. 6.21a. At equilibrium the control signal I_C and feedback signal I_F cancel, and only the triangle wave i_t exercises the window comparator. The bipolar output window comparator provides the required three output levels, positive, zero, and negative. To pro-

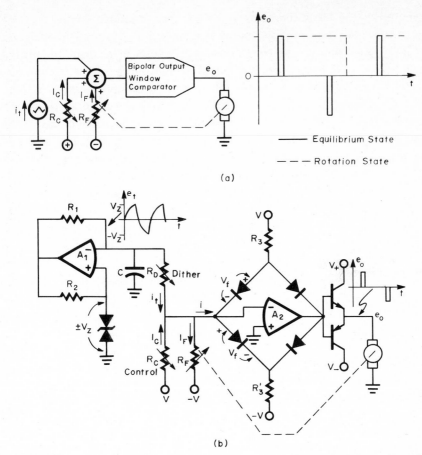

Fig. 6.21 (a) A three-level pulse width-modulated control signal with dither is derived with a triangle-wave generator and window comparator; (b) an operational amplifier realization simplifies the controller by preserving only those characteristics of the generator and comparator which are important in the controller.

vide dither pulses when exercised by i_t, the comparator trip levels are set just below the peaks of the triangle wave. To produce a shaft-angle change, the control potentiometer R_C is varied, which adds a dc offset to the triangle-wave signal. The offset holds the triangle signal above one trip point for a greater time and moves it away from the other trip point. As a result, the pulses of one polarity increase in width and the others decrease. This unbalanced output signal drives the motor, from which a change in the feedback potentiometer R_F is derived. Rotation continues until the feedback signal again balances out the control signal.

The typical triangle-wave generator and window comparator repre-

sented by the block diagrams are somewhat complex. However, each can be replaced by a simple operational amplifier circuit which preserves only those generator or comparator characteristics that are significant in the servo controller. Such a simplified controller is represented in Fig. 6.21b, where A_1 is used for the triangle-wave generator and A_2 is used for the comparator. Both amplifiers must have a high input resistance under input overload to avoid loading of surrounding elements. The triangle-wave generator is essentially that in Fig. 5.8. The triangle-wave nonlinearity makes the pulse width modulation nonlinear, but this error is removed by the null-seeking feedback loop. Similarly, the feedback loop removes window comparator errors. Moderately stable trip points ensure dither signal control, but otherwise errors are negligible under feedback. With low comparator input current i, the diode bridge around A_2 provides low impedance feedback to hold the amplifier output near zero. As signals raise the input current to the comparator above that which can be supplied by the diode bias, the output is driven away from ground level. A more detailed description of the comparator operation is given with Fig. 4.3.

6.3.3 An optimizing controller
For many systems operation is desired at an arbitrary point that results in optimum performance rather than at a fixed setpoint. Electronic control for an optimum operating point can be achieved by sweeping the control variable and using feedback to stop the sweep when the optimum is reached. This control operation is illustrated below for automatic receiver tuning. Tuning can be rapidly locked on narrowband signals in a broad frequency search band by electronically sweeping the tuned frequency until the AGC output reaches an adequate level. Such automatic tuning is provided by a feedback-controlled ramp generator, as shown in Fig. 6.22. The ramp generator is similar to the circuit of Fig. 5.17, and it uses an integrator and a comparator in a positive feedback loop to create a ramp oscillation. By using this ramp to drive a varactor tuning diode, the tuned frequency of the receiver is swept in search of a strong signal. To avoid forward-biasing the diode, the driving ramp is made zero-based. For this a zero-level comparator trip point is produced by using diode D_1 to disconnect the hysteresis feedback of the comparator on the negative swing.

When the receiver detects a signal, the AGC output rises, and this out-

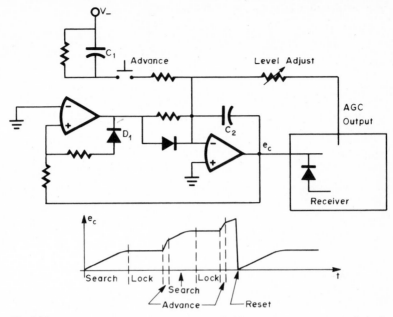

Fig. 6.22 Automatic receiver tuning is provided by a ramp generator with feedback from the AGC output.

put is fed back to control the ramp generator. At some positive level the AGC output cancels the negative integrator input signal supplied by the comparator, and integration stops. Then, the voltage on the tuning diode is held constant to lock tuning at the frequency of the detected signal. Since the integrator is in a feedback loop with the receiver, the integrator drift is compensated by the feedback. When higher-level signals are being searched for, the level adjustment is set so that only high AGC levels will stop the sweep. For very low-level signals the frequency sweep could be stopped by a peak detector, giving fine tuning to the best reception point. To remove the frequency lock and advance in search of another signal, the switch shown is depressed. This charges C_1 through an integrator summing resistor and produces an equal increase in charge on C_2. By this increase the integrator output rises to detune the receiver and thereby remove the locking AGC feedback. The permissible sweep rate of the controller is limited by the responses of the I-F strip and the AGC circuit. If the controller sweep changes tuning before a detected signal can produce an AGC output, frequency lock will not occur.

6.4 Other Special-purpose Circuits

Among the other circuits formed with operational amplifiers are audio and AGC circuits. For audio circuits operational amplifiers are well suited to the needs for stable, controllable gain and for an equalizing, variable-frequency response. High feedback-loop gain for precise AGC is also provided by operational amplifiers. Greatly improved response and linearity is achieved with phototransistor circuits by using operational amplifiers to bias the transistors and to provide linearity compensation.

6.4.1 Audio circuits   Operational amplifier circuits useful in audio applications include the low-noise amplifier of Fig. 1.27, the power circuits of Sec. 2.2, the filters of Sec. 3.5, the AGC circuits of Sec. 6.4.2, and the response control circuits of this section. Using simple active filter techniques, operational amplifiers can provide the precise response shaping needed for equalization and the response adjustment desired for tone control. Frequency response equalization is required to compensate for the varying levels at which different frequencies are commercially recorded.

A preamplifier designed for the standard RIAA equalization for records and the NAB equalization for tapes is shown in Fig. 6.23. The basic amplifier has a voltage gain of 60 dB as set by R_1 and R_2. Below 10 Hz the gain is rolled off by capacitor C_1, which adds impedance in series with R_1. At higher frequencies the switched feedback elements decrease the gain. For the RIAA equalization a response pole is developed at 50 Hz by the shunting of C_2 and C_3 on feedback resistor R_2. This shunting is interrupted by a response zero at 500 Hz produced by stop resistor R_3 in series with C_2. Then, at 2 kHz, C_3 creates a pole with this stop resistor to continue the gain rolloff. For the NAB equalization, C_4 creates a 50 Hz pole with R_2, and stop resistor R_4 produces a response zero with C_4 at 3 kHz.

As long as the operational amplifier has an open-loop response significantly greater than the desired equalizing responses, performance is essentially determined by the feedback elements. Component tolerances determine the accuracies of the response break points. Note that a polarized capacitor is shown for the large C_1 even though the signal on this capacitor is bipolar. This is possible for most preamplifier applications, as the input voltage is well below the reverse voltage limitation of the capacitor. Alternately, the capacitor can be returned to the positive

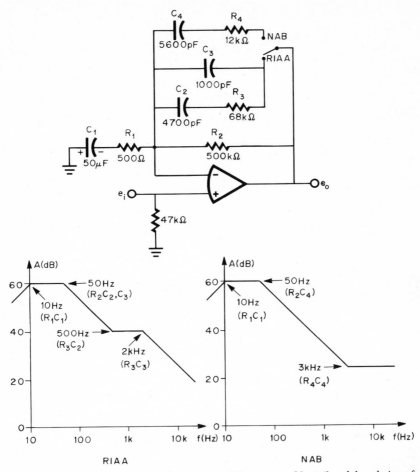

Fig 6.23 Equalization responses for a preamplifier are readily tailored by choice of operational amplifier feedback elements.

supply to avoid reversing the capacitor voltage. An input ground return resistor of 47 kΩ is shown, but this value may be altered to match the specific signal source used.

Independent control of the high- and low-frequency gains of an audio amplifier provides bass and treble adjustment. Once again, the desired response control can be achieved with simple feedback around an operational amplifier, and one such tone-control amplifier is shown in Fig. 6.24. Separate variable-feedback networks are connected for the two controls. At very low frequencies where all the capacitor impedances are high, the only effective feedback is the resistive portion of the bass

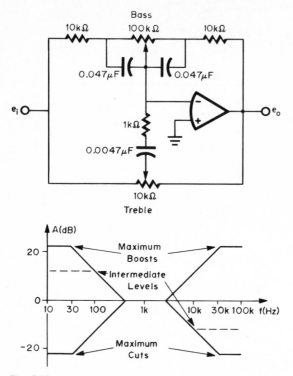

Fig. 6.24 Independent bass and treble tone control is achieved with separate feedback networks around an operational amplifier.

control. Then, the circuit acts as a simple variable-gain amplifier with a gain range of ± 22 dB at the extremes of the bass region. The effect of the bass control potentiometer is bypassed by capacitors so that the gain is held at 0 dB at somewhat higher frequencies.

At even higher frequencies the coupling capacitor of the treble control begins passing feedback signal to alter the high-frequency response. This treble feedback combines with that of the bass control, which consists simply of the 10 kΩ resistors at these frequencies. At the extremes of the treble control range the 1 kΩ resistor is effectively in parallel with one or the other of the two 10 kΩ resistors to again give a ± 22 dB control range. As either control is varied, its gain at the frequency extreme changes, but the response intercept with the 0 dB level remains fixed. For the connection shown the source resistance must be much less than 10 kΩ in order to not alter the tone control responses. Where a higher source resistance is present, the signal can drive the noninverting input of the operational amplifier with the input terminal shown grounded. While this noninverting configuration makes the midband gain 2 rather than unity, it presents the high, common-mode input impedance to the signal source. With this high impedance a capacitive coupling with smaller capacitors can also be achieved.

6.4.2 AGC circuits For amplitude stabilization under varying signal conditions an automatic gain control (AGC) is often applied. By means of feedback the gain of an amplifier is automatically controlled to maintain a constant output signal level even though the input signal varies widely. To achieve feedback control of gain a voltage-controlled gain element is needed. This control is provided by the voltage-controlled resistance of FETs and by an analog multiplier in the circuits described below. Other AGC circuits are included with the sine-wave generators of Sec. 5.1.

While an FET is a simple gain-control element, its resistance is modulated by the signal voltage on that resistance, and this modulation effect introduces distortion. In addition, a large, bipolar signal on a junction FET can forward-bias the junction to produce a drastic resistance change. To avoid these distortion effects and still permit large-signal swings, the FET is connected as the grounded element of a T network[6] in Fig. 6.25. The gain of the resulting circuit is most sensitive to modulation of the FET resistance when the resistance is low. With the T network the modulation effect is the lowest then, since the portion of the signal supported by the low FET resistance is small. At its higher resistance levels the FET supports a larger signal, but the divider action of R_2 and R_3 limits this signal to a level that does not forward-bias the gate.

Through the variable resistance of the FET, the gain of the amplifier of Fig. 6.25 is controlled to maintain a nearly constant average output level. As the FET resistance varies from a minimum of r_{ON} to a maximum of r_{OFF}, the gain ranges over

$$-\frac{R_2 + R_3}{R_1} \geq A \geq -\frac{R_2 + R_3}{R_1} - \frac{R_2 R_3}{r_{ON} R_1}$$

where r_{OFF} is assumed to be negligibly large. A gain range of 40 dB is attainable. To control this gain the output amplitude is detected by the diodes shown, and the detected signal is averaged by a capacitor for sus-

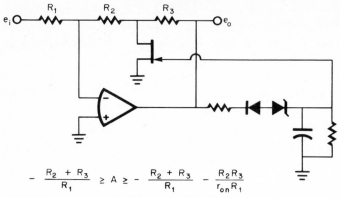

$$-\frac{R_2 + R_3}{R_1} \geq A \geq -\frac{R_2 + R_3}{R_1} - \frac{R_2 R_3}{r_{on} R_1}$$

Fig. 6.25   AGC control with an FET has greatly reduced distortion when the FET is connected in a T network to reduce the signal swing on the FET.

tained gate bias. If the amplitude falls below the detector level, no gate bias is developed, and the FET assumes its minimum resistance. This in turn raises the gain to restore the output amplitude. Increased amplitude is detected and results in a gate bias that increases resistance until equilibrium is reached. At equilibrium the FET resistance, amplifier gain, output amplitude, and gate bias are at mutually compatible levels.

While the zener diode is the major determinant of the equilibrium amplitude, the FET and its bias also have effects. The output amplitude needed to produce the gate bias varies with the input signal. A change in this bias is needed to adjust gain when the input signal varies. As a result, the amplitude of the output still varies with that of the input, but this variation is greatly reduced by the AGC feedback. Another amplitude variation results from changes in signal frequency. This is due to the associated change in ripple voltage on the filter capacitor.

Greater amplitude stability can be achieved with an AGC circuit that has a more ideal detector diode and that has gain in the control feedback. With an ideal diode no signal would be lost in supporting a diode voltage drop, and so very small amplitudes could be detected. Gain is needed to reduce the signal swing required to develop the AGC feedback signal. These improvements are shown with a basic AGC amplifier in Fig. 6.26. In this circuit the amplifier formed with A_1 receives control feedback from an integrating absolute-value circuit.[7] From the precision rectification a near-ideal detection is achieved, and the high gain of the integrator develops the bias for the AGC FET from only a very small portion of the output signal. In addition, the integrator provides the filtering needed to sustain the AGC bias between signal peaks.

Operation of the absolute-value circuit is similar to that of Fig. 4.10 except for the integrating operation. As before, the signals summed at the input of A_3 have a constant polarity sum. The difference between this sum and the reference E_R is the signal integrated, and the average of this difference produces a change in the integrator output to adjust gain. The gain adjustment will be in a direction which will reduce the average of this difference signal to zero. At this point equilibrium exists, and the integrator holds the FET bias at a fixed level. Equilibrium exists when the average of the output signal magnitude equals the reference, as expressed by

$$|e_o|_{\text{avg}} = E_R$$

To reduce the distortion introduced by AGC, the gain control element can be made an analog multiplier rather than an FET. This avoids the distortion introduced by signal modulation of the FET resistance and results in a distortion level determined primarily by the linearity of the multiplier. For this approach the amplitude detection and feedback-control voltage can be provided by the integrating absolute-value circuit

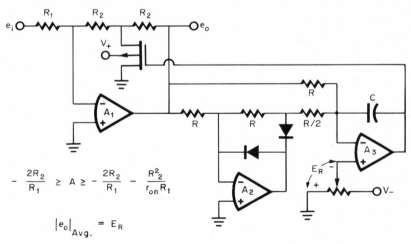

$$-\frac{2R_2}{R_1} \geq A \geq -\frac{2R_2}{R_1} - \frac{R_2^2}{r_{on}R_1}$$

$$|e_o|_{\text{Avg.}} = E_R$$

Fig. 6.26 An integrating precision rectifier removes detector and feedback errors from AGC.

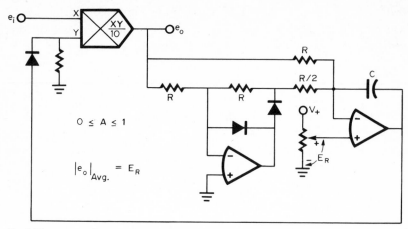

Fig. 6.27 AGC with an analog multiplier can be achieved with feedback through an integrating absolute-value circuit.

of the last figure. The resulting circuit combination is shown in Fig. 6.27. Once again, the precision rectification provides near-ideal amplitude detection without voltage loss for diode drops, and the integration develops a feedback voltage that stabilizes when the average of the rectified signal equals the reference voltage E_R. The integrator output determines a multiplier input voltage to set the gain by which the input signal is multiplied. As before, the feedback stabilizes the average amplitude at a level that equals the reference voltage, or

$$|e_o|_{avg} = E_R$$

REFERENCES

1. J. G. Graeme, Accurate Transistor Tests Can Be Made Inexpensively, *Electronics,* Feb. 28, 1972.
2. G. E. Tobey, J. G. Graeme, and L. P. Huelsman, *Operational Amplifiers: Design and Applications,* McGraw-Hill Book Company, New York, 1971.
3. *Applications Manual for Operational Amplifiers,* Philbrick/Nexus Research, Dedham, Mass., 1965.
4. T. Kohler and E. Hudspeth, Solid State Op Amp Measures Kilovolts with 0.5% Accuracy, *Electronics,* Dec. 25, 1967.
5. J. G. Graeme, Using Operational Amplifiers for DC Servo Control, *Instrum. and Control Syst.,* April 1972.
6. L. Accardi, FET Gain Control Stage Handles High Level Signals, *Electron. Design,* March 2, 1972.
7. E. Guenther, MOSFET Provides 60dB Dynamic Range Low-Frequency AGC Circuit, *EEE,* November 1969.

GLOSSARY

Absolute-value Circuit   A circuit that produces a unipolar output signal equal to the magnitude or absolute value of a bipolar input signal.

Acquisition Time   The transition time required by a sample-hold circuit to switch from the HOLD mode to the SAMPLE mode.

Aperture Time   The transition time required by a sample-hold circuit to switch from the SAMPLE mode to the HOLD mode.

Bandwidth   See Unity-gain Bandwidth; Full-power Response.

Bias Current   See Input Bias Current.

Bode Diagram   A straight-line approximation to a gain-magnitude- or phase-response curve.

Breakpoint   A point at which a Bode diagram changes slope due to a response pole or zero at that frequency.

Broadbanding   A phase compensation technique in which the compensation applied to an amplifier is reduced for broader bandwidths at higher closed-loop gain levels.

Charge Amplifier   An amplifier that produces an output voltage proportional to the charge supplied to its input.

Chopper-stabilized Amplifier   An amplifier stabilized against dc drift by chopping the input signal to form an ac waveform that can be coupled through dc isolating capacitors to an amplifier and later demodulated.

Common-mode Input Capacitance, C_{Icm} The effective capacitance between either input of an operational amplifier and the common.

Common-mode Input Resistance, R_{Icm} The effective resistance between either input of an operational amplifier and the common.

Common-mode Voltage   The average of the two voltages applied to differential amplifier inputs.

Common-mode Rejection Ratio (CMRR)   The ratio of the differential voltage gain of an amplifier to its common-mode voltage gain.

Comparator   A differential input amplifier used to compare the voltage levels at its two inputs and having high gain so that only small voltage differences are needed to switch the output voltage from one polarity to the other.

Current Limiting   A means for limiting the output current supplied by an amplifier for protection purposes.

Difference Amplifier   An operational amplifier with a feedback configuration that results in an output voltage proportional to the difference of two input voltages.

Differential Input Amplifier   An amplifier which has two inputs of opposite gain polarity with respect to the output.

Differential Input Capacitance, C_I The effective capacitance between the two inputs of an operational amplifier when operated open loop.

Differential Input Resistance, R_I The effective resistance between the two inputs of an operational amplifier when operated open loop.

Differential Output Amplifier   An amplifier which has two outputs of opposite gain polarity with respect to a given input.

Differentiator   An operational amplifier with a feedback configuration that results in an output signal proportional to the time derivative of the input signal.

Drift   See Input Bias Current Drift; Input Offset Current Drift; Input Offset Voltage Drift.

Feedback   The return of a portion of the output signal from a device to the input of the device.

Feedback Factor, β That fraction of an output signal fed back to the input.

Feedforward   An amplifier phase compensation technique in which high-frequency signals are fed forward around the low-frequency portion of the amplifier.

Full-power Response, f_p The maximum frequency at which an operational amplifier can supply its rated output voltage and current without significant distortion.

Function Generator   A circuit that produces an output signal voltage related to an input signal by some specific or adjustable function.

Frequency Compensation   See Phase Compensation.

Frequency Response   See Unity-gain Bandwidth; Full-power Response.

Gain   See Open-loop Gain; Loop Gain.

Gain Error   The difference between the actual closed-loop gain of an operational amplifier with feedback and that predicted by the ideal gain expression.

Guarding   See Input Guarding.

Hysteresis   The transfer response lag of comparators controlled by positive feedback and resulting in different trip points for the two directions of output transition.

Input Bias Current, I_B The dc biasing current required at each input of an operational amplifier to provide zero output voltage when the signal and input offset voltages are zero.

Input Bias Current Drift   The rate of change of input bias current with temperature or time.

Input Capacitance   See Common-mode Input Capacitance; Differential Input Capacitance.

Input Guarding   Use of an input shield that is sometimes driven to follow the voltage level of the input signal and, thereby, remove leakage and loss-inducing voltage differences between the input path and surrounding stray conduction paths.

Input Noise Current, i_n The equivalent input noise current at each input of an operational amplifier which would reproduce the output noise if the various sources of amplifier current noise were set to zero and if the input noise voltage were zero.

Input Noise Voltage, e_n The equivalent differential input noise voltage of an operational amplifier which would reproduce the output noise if the various sources of amplifier voltage noise were set to zero and if the input noise current were zero.

Input Offset Current, I_{OS} The difference between the two input bias currents of a differential input operational amplifier.

Input Offset Current Drift   The rate of change of input offset current with temperature or time.

Input Offset Voltage, V_{OS} The dc input voltage required to provide zero voltage at the output of an operational amplifier when the input bias current is zero.

Input Offset Voltage Drift   The rate of change of input offset voltage with temperature or time.

Input Protection   Protection of the input of a device from damage due to the application of excessive input voltage.

Input Resistance   See Common-mode Input Resistance; Differential Input Resistance.

Instrumentation Amplifier   A direct-coupled, differential input amplifier with internal feedback committed for voltage gain.

Integrator   An operational amplifier with a feedback configuration that results in an output signal proportional to the time integral of the input signal.

Integrator Reset   Charging the integrator capacitor to a specific dc level to establish the initial condition of the integration.

Inverting Amplifier   An operational amplifier with a feedback configuration that results in negative voltage gain and, thereby, inversion of the signal polarity.

Isolation Amplifier   An amplifier with high-impedance, high-voltage isolation between input and output commons.

Logarithmic Amplifier   An amplifier which develops an output voltage that is proportional to the logarithm of the input signal.

Loop Gain, Aβ   The gain around a feedback loop formed by an amplifier and its feedback network.

Noise   See Input Noise Current; Input Noise Voltage.

Noninverting Amplifier   An operational amplifier with feedback connected for positive voltage gain, which thereby does not invert the signal polarity.

Offset Current   See Input Offset Current.

Offset Voltage   See Input Offset Voltage.

Open-loop Gain, A   The ratio of the output signal voltage of an operational amplifier to the associated input signal voltage when the feedback loop is open-circuited.

Operational Amplifier   A high-gain dc voltage amplifier which has high input impedance and low output impedance and is capable of developing bipolar output signals from bipolar input signals.

Output Protection   Protection of the output of a device from overloads, as commonly provided by output current limiting for an operational amplifier.

Overload Recovery Time   The time required for the output of an operational amplifier to return from saturation to linear operation following the removal of an input overdrive signal.

Phase Compensation   Frequency response tailoring for stability of a feedback system through addition of response poles and zeroes that reduce high-frequency phase shift.

Phase Margin, ϕ_m   The margin by which phase shift around a feedback loop is less than 360° at the unity-loop-gain frequency.   Sometimes defined as the margin by which the phase lag of a feedback amplifier is less than 180°, for the negative feedback accounts for the other 180° of phase shift.

Power Booster   A buffer amplifier with high current gain and typically unity voltage gain, used to boost the output power from an operational amplifier.

Precision Rectifier   See Absolute-value Circuit.

RMS Converter   A circuit that develops a dc output voltage equal in rms value to an input signal of arbitrary waveform.

Sample-hold Circuit   A device whose output follows an input signal and then holds the instantaneous value of the signal that existed when the HOLD command signal was applied.

Settling Time, t_s   The time required for output voltage to settle to within a specified percentage of its final value following a step input.

Signal Conditioner   A device that conditions or modifies a signal so as to alter the relationship of the signal with respect to time, frequency, or other signals.

Signal Processor   A device that acts on or converts signals by analyzing, routing, rectifying, sampling, etc.

Single-ended   Characterized by a single input or output rather than by the two of a differential input or output.

Slewing Rate, S_r   The maximum rate of change of an output voltage when supplying the rated output.

Summing Junction   The junction of the feedback and input resistors of a feedback network at which the signal currents from input resistors are summed.

Unity-gain Bandwidth, f_c The frequency at which the open-loop gain of an operational amplifier crosses unity.

Varactor Amplifier   A modulated-carrier dc amplifier using the capacitance modulation of varactor diodes in response to low-frequency signals to transmit the signal for ac rather than dc amplification.

Virtual Ground   A characteristic of the summing junction of an inverting connected operational amplifier, which resides virtually at ground potential since the very high open-loop gain of the amplifier requires only small summing junction signals to develop the output signal.

Voltage Follower   The short-circuit feedback connection of an operational amplifier which results in an output signal that follows the voltage at the noninverting amplifier input.

Window Comparator   A comparator that detect levels within a set range or window rather than simply distinguishing between levels above and below a set point.

INDEX

INDEX

229